Audio Augmented Reality

Audio Augmented Reality: Concepts, Technologies and Narratives provides readers with a comprehensive overview of audio augmented reality (AAR), focussing on its narrative potential while discussing several design considerations and prospective application domains.

In this groundbreaking book, sound designer Matias Harju provides a practical and insightful exploration of the medium. The book draws perspectives from sound designers, researchers, and industry professionals who are actively shaping this rapidly evolving field.

Designed to inspire and offer practical insights, *Audio Augmented Reality: Concepts, Technologies and Narratives* serves as a guide for creators, academics, and anyone interested in exploring the creative potential of AAR.

Matias Harju is a Helsinki-based sound designer and musician specialising in immersive and interactive sonic storytelling. His creative practice centres on exploring audio augmented reality as a narrative art form, while he also works on virtual reality, video games, installations, and live performances. With a diverse background in music, audio, and education, he holds Master's degrees from Sibelius Academy and the Sound in New Media programme at Aalto University. Matias regularly lectures and leads workshops on spatial sound design and audio augmented reality.

Audio Augmented Reality

Concepts, Technologies and Narratives

Matias Harju

Routledge
Taylor & Francis Group

LONDON AND NEW YORK

Designed cover image: Shutterstock

First published 2025
by Routledge
4 Park Square, Milton Park, Abingdon, Oxon OX14 4RN

and by Routledge
605 Third Avenue, New York, NY 10158

Routledge is an imprint of the Taylor & Francis Group, an informa business

British Library Cataloguing-in-Publication Data
A catalogue record for this book is available from the British Library

ISBN: 978-1-032-79202-6 (hbk)
ISBN: 978-1-041-04197-9 (pbk)
ISBN: 978-1-003-62728-9 (ebk)

DOI: 10.4324/9781003627289

Typeset in Sabon
by Taylor & Francis Books

Contents

Acknowledgements

First and foremost, I want to thank Ville Walo of WHS, without whom the Full-AAR project would never have started in 2021, eventually leading to the writing of this book. I'm also grateful to the rest of the Full-AAR team, people at WHS, and the actors, musicians, sound engineers, and test participants who have contributed to the project.

I want to express my sincere thanks to Evie Evans at Routledge, who suggested this book project in autumn 2023 and has been a great support since. Thanks also to other awesome people at Routledge and copyeditor Kristina Wischenkamper for working with me to turn my thoughts into a book.

A big thank you goes to Professor Iain McGregor for both invaluable feedback on the initial book proposal as well as thorough and wise comments on the draft manuscript. My thanks also go to the other pre-reviewers, Professor Sebastian Schlecht, and Professor Stefania Serafin.

I am grateful to everyone with whom I have had the honour of discussing and who have shared invaluable insights. Special thanks go to Thomas Aichinger, Steffen Armbruster, Rémi Audfray, Professor Karlheinz Brandenburg, Halsey Burgund, Professor Michael Cohen, Holger Förterer, JiaYing Grygiel, Hanna Haaslahti, Edwin van der Heide, Mikko Honkanen, Miranda Kastemaa, Josh Kopeček, Mari Koppinen, Nils Meyer-Kahlen, Hanna Niittymäki, Mathis Nitschke, Ronja Pahaoja, Professor Ville Pulkki, and Ulrike Sloma.

I want to thank Emilia Lehtinen, author and filmmaker, for brainstorming invaluable narrative ideas for AAR during the past years. Thank you also, Jussi Ratilainen, for the captivating cover picture.

Finally, my deepest thanks to you, Kaisa, my wife, for your unwavering support and for making sure I had the time to focus when it was needed most.

1 Introduction

Year 2045

'Let's keep it audio-only', said one of the voices at the start.

Running online meetings with cameras off had become almost the norm, like a reminiscence from the pandemic years of the early 2020s. I didn't mind, since I felt that I could concentrate better on the conversation when only listening. Thanks to the clear-cut audio quality and latency no bigger than the speed of light, I was able to read emotions and intentions in people's voices and reactions. The best feature of the current telepresence applications was, however, the spatialisation: remote colleagues' voices were positioned around me as authentically as if they were in the same room. This made it easy to follow conversations, even when people spoke simultaneously. I could also move closer or walk around my virtual peers. At one point, when two colleagues wanted a private exchange, their voices shifted a few metres away to avoid disturbing the group, still staying within earshot to rejoin the discussion seamlessly.

After my part was covered, I said my goodbyes and hung up the call by moving my eyes in a pre-defined pattern; the earbuds detected and interpreted the small sounds caused by the eye movements. I was left with the hum of the air conditioning.

'They can put a man on Mars, but they can't make A/C any quieter...' I muttered, although I did not actually complain. The earbuds were technically better than able to detect and erase the hum, but somehow I felt the noise soothing and let it pass to my ears.

After work, I bicycled to a friend of mine who had just moved to the area. The electronic audio pass-through was so transparent and the binaural soundscape so realistic that I felt safe in traffic even with the buds on. A colleague of mine joined me for the first two kilometres. We tried to talk, but the traffic noise made it hard to catch more than a

DOI: 10.4324/9781003627289-1

word here and there. So, we switched our earbuds to pairing mode. Now, the earbuds with their internal microphones acted like wireless transceivers, letting us talk like motorcyclists with their helmet radios. But just like in the online meeting, I could hear my colleague's voice spatialised to where he was.

My colleague and I continued along different paths. As I was uncertain about the last turns to take, the navigation app started to give me directions by playing a beacon sound spatialised always at the next waypoint, letting me intuitively choose the right exits in intersections. I also received a few warning sounds when a fast cyclist was passing me from behind without ringing the bell. I did hear a bell, though, but that was generated by my earbuds and spatialised in the direction and distance of the fellow rouleur.

I was standing in an alley leading to an urban district of newly built apartment houses. The street areas between buildings were still under construction. On my right, there was a small patch of trees, and a dull wave of traffic noise was pouring over the grove. Behind the trees, I could see the face of a noise barrier, obviously incapable of performing its duty.

'Hey buds, erase traffic', I said softly. In a second, the traffic noise disappeared, unmasking the sound of birds singing in the trees. Birds seem to adapt to any environment, not unlike humans, but without any technological aid. I could have set the earbuds to automatically erase any distant traffic noise by default, but as that would limit my understanding of the surrounding world, I preferred manual selection.

My friend appeared from the direction of the houses.

'Those birds sound happy', I greeted her.

Her face lit up. 'Well, they should be as we have some nice plans for this neighbourhood. Currently, the street canyons look and sound ghostly without any green whatsoever anywhere, but there are large tree beds reserved under the streets, planting strips for wall climbers next to each house, and the roofs will also be growing plants.'

'The only thing missing is money, I guess', I remarked.

'You guess right, but we may have something to help to fix that. Do you want to hear what this place would sound like if the plans get realised?'

She operated her smartwatch and looked at me with a question on her face. I nodded, giving her my silent permission to receive an audio feed. I heard a soft voice in my ear asking whether I wanted to accept her incoming audio feed. I nodded, and after some seconds the few birds were greeted by dozens of their friends and rivals, surrounding me from all directions. The reverberant voices of the children playing soccer on the nearby field between large buildings got softer as the echoes got dampened by the invisible vegetation.

We started to walk towards the buildings and dove into one of the street canyons. The virtual nature sounds evolved naturally along the path: trees and vines were rustling, squirrels were running from tree to tree. We saw a construction worker drop a metal bar on the dusty pavement but heard just a soft thump. What I learned later was that the system had been measuring my heart rate and adjusting the pace of the sounds to make me feel even more relaxed.

'What we're aiming at now is to make authorities hear these sounds-capes on location. The sounds will make them feel this place. They will understand the need for biophilic design. Finally, I hope, they'll loosen their purse strings.'

After the audio walk, we visited a home museum nearby. It used to belong to a local artist and his wife, a professional composer who had been left in the shadows of her husband. Since we had compatible earbuds, we accepted the audio feed from the museum's media server. For those without personal devices, the museum borrowed dedicated headphones.

I heard someone play the piano in the living room. While walking there, I spotted a grand piano in the middle of the room, however, no one was playing it. Still, the music was emanating from the instrument, sounding very realistic and changing its timbre when moving around the large body of the piano. My friend was drawn by the paintings on the wall, not paying attention to the instrument.

The music was fragmented as if someone was trying out different variations of a musical idea. Soon I heard a voice, a woman's voice, yelling from where the pianist would be sitting.

'Dear, what do you think, is this harmonic progression too obvious?!'

A male voice replied from somewhere else in the house, perhaps the kitchen since I could hear sounds of cutlery and other kitchen items in the background.

'Go for it! To me, it sounds fresh.'

There was hand-written sheet music on the music stand. As I went closer to look at it, the system understood that I was interested in the composition and let the immaterial woman continue working on the piece, humming faintly and trying out harmonies on the piano. I saw my friend walk to the kitchen: she must have been following the artist's story.

The exhibition continued in a similar manner. Sometimes, the sound sources were outside, like patter of rain on the roof, or in other rooms, like a quarrel in the upstairs bedroom. The fact that we could not see the sources made the sounds feel strangely real. Other times, the sounds were floating in the rooms as ghosts of the family that once lived in the house, still sounding realistic had we closed our eyes or been blind. The sound experience did not just make the home feel alive but made us better understand the two divergent narratives of the house dwellers.

Later in the evening, we went for dinner. The restaurant was packed, and happy people talking over each other managed to generate a huge noise. I immediately activated the speech separation mode. The restaurant hubbub faded out, but when a waiter came to greet us, in a few seconds, my earbuds had analysed his voice and let it pass isolated and clear. In a minute, my friend's neighbour joined us, and her voice was also isolated to enable relaxed conversation without the need to fight the volume war.

However, we had difficulties in finding a common language that each of us spoke fluently to carry on a comfortable conversation. To solve the problem, we activated the interpretation mode. The earbuds then listened to the neighbour's speech and translated that in real time. The system masked her original voice, replaced that with a speech synthesis in our language, and spatialised it as if coming from her mouth. There was no lip-sync, of course, but the virtual voice's tonal quality was so close to her own, and the virtual acoustic simulation managed to fit the voice perfectly in the current environment, making it easy to forget the wizardry that was happening in the microprocessors.

The evening air felt gentle as we set off, and instead of hopping on my bike, I chose to stroll. I had been engrossed in a spy audio game for the past few days and wanted to keep unravelling the mystery during my walk home.

'Buds, spy game.'

Before long, I heard a 'Pssst!' from a nearby alley. Following the sound, I could hear a secret message from my informant about enemy spies having a meeting in a brick building close by, and I was encouraged to eavesdrop. The message had been left by another player—not in that particular alley, but in an alley somewhere in their own neighbourhood. The game running in my earbuds had analysed my surroundings and identified a similar environmental feature to place the message. This way, the story became local for everyone.

After a while, I indeed spotted an abandoned brick storage house and walked closer. Through one of the windows, I could hear a fictional conversation unfolding inside.

'...and we'll keep all the communication encrypted all the time. The prime minister's right hand will photograph the secret memorandum and pass it to us. We have to act fast before the paper gets to wider circulation...'

So the prime minister's aid is a spy! I continued my walk, prepared a voice message, and 'dropped' it behind the next bus stop. In another city, sometime after, another player would find my message behind a bus stop and learn a detail about the conspiracy happening in the prime minister's cabinet.

Introduction

Hearing is our most important sense. While we may live in a visually dominated culture, one can argue that it is through ears that we comprehend the world. Unlike vision, hearing is omnidirectional, 360 degrees to all directions, and able to detect and recognise sources behind obstacles. It also works in darkness and even when we sleep. In the absence of hearing, the sense of presence and understanding of the environment suffers greatly—the world seems less alive (McGill et al., 2020). While eyes may offer us more precise data, ears connect us three-dimensionally and emotionally to the world.

Audio augmented reality (AAR) merges virtual sounds with the real environment. With AAR, the world speaks to us in a way we have previously not been able to hear. When information and narratives are attached to the environment and objects in the form of sounds, we can start relying on our intuition, feel the world more deeply. AAR is able to create the illusion that the virtual sounds coexist with the reality: their direction and distance feel correct, and they acoustically fit in with the environment. When these sounds start to react to the real world, including the user, they feel even more alive.

Moreover, sensory augmentation offers us the ability to hear real-world phenomena that are out of ear's range—too quiet, too far away, too low, too high—or not sounds at all such as radiation, distance and temperature variations. Not only can AAR add sounds, it can also remove real ones. Manipulation of personal soundscapes to remove unwanted sounds and enhance wanted ones offers intriguing possibilities for uplifting the quality of life. Some state-of-the-art noise cancelling headphones and medical hearing aids are already able to recognise speech and other sounds and isolate them from the background noise. More ways to control how we hear the world around us are coming, helping social interaction and saving our hearing for precious unmediated moments.

AAR will also be a key technology when connecting our lives to digital ecosystems. In Lev Manovich's (2003) terms, we are living in an 'augmented space' where real-time information, synchronised with online data, is constantly available in multiple interfaces around us and carried by us. While wearable 'smart' devices such as watches and rings represent the discreet end of interfaces to this world, *augmented reality* (AR) glasses and *mixed reality* (MR) headsets appear as media-sexy portals to the digital life. Yet, while they allow hands-free operation with graphical information projected in front of our eyes, such devices are either bulky, or the image quality is poor—or both. Wearable AAR devices, on the other hand, in the form of headphones, earbuds, and smart glasses with near-ear speakers, are relatively comfortable and discreet, and already socially accepted. While their capability to render 3D sound is not yet perfect, the technology is still less complex to integrate in small form factors compared to visual display systems.

As an audio-first approach, AAR is likely to find its ecological pocket in this jungle of *metaverse, pervasive technology, ubiquitous computing, internet of things* and *spatial computing*. As we have already seen since the release of Sony Walkman in 1979 and later with the huge popularity of podcasts and audio books, personal audio has become a convenient and natural way to engage with information, entertainment, and narratives. With interactive applications, compared to looking at the mobile phone screen, audio interfaces can pull 'people out of their screens so that we can be heads up, hands free, and more IN the world around us' (Bye, 2019).

This book discusses AAR as a medium, technology, and art form. Instead of approaching it as a listening experience, the attempt is to understand it from the perspective of reality: how AAR can be used to manipulate our perception or augment the real world through acoustic layers conveying data and narratives.

AAR is a multi-disciplinary medium, drawing upon science, technology, and art. While the book attempts to be an overall introduction to this medium and shine light from various angles, it has a bias towards AAR's narrative use and possibilities. That is an intentional framing to remind that AAR is not just technology, but an intriguing possibility to tell stories. Humans comprehend the world, in the end, through stories. Even the most productivity-oriented AAR applications can benefit from narratives and emotional content.

The first chapter of the book offers an overview of AAR with characteristic concepts and approaches, including AAR in relation to AR, virtual sounds, plausibility, various auditory displays, and acousmatic sounds. Chapter 2 discusses the concepts of reality, sense of presence, interactivity, and relationship with the environment. Chapters 4 and 5 present some key historical events influencing the development of the medium, while the main focus is on applications and experiences that have been pioneering or are otherwise representative of the diverse forms of AAR.

Chapter 6 discusses the basic psychoacoustic concepts and the principles of the virtual audio engine that makes audio augmentations possible. Chapter 7 concentrates on the technical components of binaural, wearable AAR.

The last part, Chapters 8 and 9, concentrate on AAR as a narrative medium. First, various narrative design considerations are discussed. Then, a framework of narrative techniques characteristic of AAR is presented. As this type of work has rarely been undertaken, the aim is to identify some of these concepts and provide users with tools to analyse AAR experiences while offering practitioners building blocks and ideas for their own creative work.

This book started with a vision for the everyday use of personal, ubiquitous AAR. While the earbuds described in the story do not yet exist at the time of writing, and the story remains science fiction, each individual feature and capability already exists at various stages of development. Some are in the form of early prototypes and proofs-of-concept within university labs or commercial research departments, while others have already reached the public as products or services. The final chapter paints a future of AAR and briefly discusses the current development status of each feature mentioned in the story. It is entirely possible that all these capabilities could soon be integrated into the form factor of earbuds, particularly if paired with a mobile phone for offloading certain tasks. However, realising this vision will depend on funders and technology companies recognising a significant market demand for the manipulation of personal acoustic environments. Equally important, society must engage in a thoughtful dialogue about the challenges and broader societal risks that AAR, alongside other wearable and ubiquitous systems, might introduce. These include the risk of personal behavioural data, collected by various sensors, being shared with third parties, as well as the threat of digital deception through deepfake and manipulated audio content (Turner, 2022).

Although the concepts behind AAR are old, and the first computer-generated AAR experiments were conducted already in the 1990s (Cohen et al., 1993; Bederson, 1995), AAR has mostly stayed within the realm of academic interest, primarily because the technology has not yet matured to support it. On the other hand, features that can be labelled as AAR have been around for a long time, starting from the use of hidden loudspeakers in attractions to headphone-based interactive museum audio guides. There are geolocated audio walks, 3D audio displays in aviation, electric cars that sound like spaceships, bat detectors, headphones that separate speech from noise... This book attempts to bring together all of these and many other phenomena within the common framework of AAR. The goal is to comprehend what has been accomplished in this field and how these diverse concepts can be combined to improve our lives, help us communicate better with each other, and deepen our understanding of the world around us.

References

Bederson, B.B. (1995) 'Audio Augmented Reality: A Prototype Automated Tour Guide', in *Conference Companion on Human Factors in Computing Systems – CHI '95*. Denver, Colorado, United States: ACM Press, pp. 210–211. Available at: https://doi.org/10.1145/223355.223526.

Bye, K. (2019) '#753: AR Spatial Audio on Bose AR Frames & QuietComfort 35 Headphones – Voices of VR Podcast', 8 April. Available at: https://voicesofvr.com/753-ar-spatial-audio-on-bose-ar-frames-quietcomfort-35-headphones/ (Accessed: 29 September 2024).

Cohen, M., Aoki, S. and Koizumi, N. (1993) 'Augmented Audio Reality: Telepresence/VR Hybrid Acoustic Environments', in *Proceedings of 1993 2nd IEEE International Workshop on Robot and Human Communication*. Tokyo, Japan, pp. 361–364. Available at: https://doi.org/10.1109/ROMAN.1993.367692.

Manovich, L. (2003) 'The Poetics of Augmented Space', in A. Everett and J.T. Caldwell (eds) *New Media: Theories and Practices of Digitextuality*. Routledge, pp. 75–92.

McGill, M. et al. (2020) 'Acoustic Transparency and the Changing Soundscape of Auditory Mixed Reality', in *Proceedings of the 2020 CHI Conference on Human Factors in Computing Systems*. Honolulu, Hawaii, USA: ACM Press, pp. 1–16. Available at: https://doi.org/10.1145/3313831.3376702.

Turner, C. (2022) 'Augmented Reality, Augmented Epistemology, and the Real-World Web', *Philosophy & Technology*, 35 (1), p. 19. Available at: https://doi.org/10.1007/s13347-022-00496-5.

2 Nature of AAR

2.1 Audio augmented reality

Audio augmented reality (AAR) enhances reality with virtual sounds. In its 'purest' form, these sounds are embedded in the environment, conveying meaningful information about it. AAR can also be used to erase unwanted real-world sounds from the user's perception. Additionally, hearing enhancement applications, such as isolating speech from background noise, are often regarded as a form of AAR. Therefore, rather than being a single technology or medium, AAR serves as a framework encompassing a wide range of possibilities for enhancing and reshaping our perception of reality.

In one example of an AAR application, the user wears headphones. A computer system plays back voices and other sounds while attempting to spatialise them so that the virtual sounds appear as coexisting with the real world. The content could range from navigational directions in a foreign city to a historical story told through sounds and voices embedded in the environment around the user.

Instead of using headphones, a more straightforward solution to embed virtual sounds in the real environment would be to conceal loudspeakers inside or near objects, a widely used trick with its roots in the ancient history (see Chapter 4.1). There are also other headphone-free approaches that would potentially enable AAR such as directional speakers, wave field synthesis (WFS), cross-talk cancellation (CTC), compensated amplitude panning (CAP), and wearable speakers, although examples of AAR applications using these methods are either theoretical or hard to find.

The definition of AAR is still in flux with different understandings of what can be considered as audio AR. Some require that an AAR application should present virtual sounds so that they are spatially fixed with the environment as if they were naturally embedded in there (e.g.,

DOI: 10.4324/9781003627289-2

Figure 2.1 A participant listening to virtual dialogue taking place behind a door in *The Reign Union* AAR story
Source: Photograph by Jussi Ratilainen

Tikander, 2009). Some others have a much more permissive take: for them, the general definition of AAR embraces any system that overlays artificial or enhanced sound within reality, for example, public announcement systems, the telephone, guitar effects pedals, and even karaoke (Mariette, 2013; Krzyzaniak, Frohlich and Jackson, 2019).

To grasp all these views, we can consider AAR as a continuum where the real world without any augmentations smoothly transitions into AAR as artificial sound elements or modifications are introduced, leading to a seamless mix of real and virtual. Thus, there would be no clear definite point when a phenomenon becomes AAR (see Figure 2.2). This would be in line with the classic *reality-virtuality continuum* (Milgram and Kishino, 1994) discussed later.

However, even though karaoke would be a fascinating topic to discuss, this book concentrates on two types of AAR applications leaning towards the right end of the aforementioned continuum.

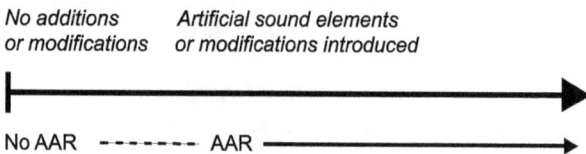

Figure 2.2 AAR continuum

1   The first type of applications attempt to create an illusion of the sounds coexisting with the reality. The virtual sounds would be spatially embedded in the real environment and convey meaningful information about it. The sounds can either appear realistic and create an illusion where the user is unaware of the mediation happening (Larsson et al., 2010), or they can be clearly distinguishable from the real ones for informational clarity or stylistic reasons (Härmä et al., 2003; Dam et al., 2024). These applications can be called mixed reality (MR) due to their high degree of *local presence*, or sense that the virtual objects are part of the reality (Verhagen et al., 2014; Rauschnabel et al., 2022).

2   The second type includes wearable applications where the audio content is closely tied to reality, but the virtual sounds are not necessarily embedded three-dimensionally in the environment. An example would be a personal, *geolocated* tourist audio guide application where voice descriptions of points of interest are played back through headphones based on the user's location in the city. Compared to, say, train announcements which are also triggered based on the current location, the personalised and often embodied experience of headphone-based audio guides create, at best, a very immersive experience.

Despite the content, a common goal for most AAR platforms is to make the technology intuitive and invisible to the user (Wellner, 2020); instead of watching a mobile phone screen or peeking through a headset display, the sounds—virtual and real—are just there, out in the world to be heard.

2.2 Augmented reality

AAR is a subset of augmented reality (AR). In theory, AR encompasses all sensory channels (Azuma et al., 2001; Schmalstieg and Hollerer, 2016), allowing augmentations through visual, auditory, tactile, olfactory, and why not gustatory, vestibular and proprioceptive senses, and in any combination of these. In practice, however, visual augmentations are by far the most common with audio coming next, leaving other sensory augmentations as special cases, such as tactile and vestibular augmentation in an earthquake simulator. When talking about *audio* AR, we usually refer to applications that manipulate just hearing and no other senses. Graphical and tactile user interfaces may be used, but they are seldom using any AR features on their own.

Augmented reality is 'grounded in reality' (Jacuzzi, 2018, p. 1); the content is directly related to the real world, and the user stays open to their surroundings. This active relationship and merging with the reality makes AR fundamentally different from *virtual reality* (VR) that aims at shutting down the user's perception of the surroundings in order to replace it with an alternative reality. For example, audio VR games are being developed for visually impaired people (e.g., Aalborg University, 2024), which are detached from the reality with the aim of immersing the user inside a virtual acoustic world. In contrast, AR registers or combines information from both physical and virtual worlds simultaneously (Schraffenberger, 2018).

2.2.1 Azuma's AR

Robert Azuma (1997) formulated a widely agreed definition of AR, with updates a few years later (Azuma et al., 2001); this states that '[a]n AR system supplements the real world with virtual (computer-generated) objects that appear to coexist in the same space as the real world' (p. 34). Azuma and his colleagues set three properties for AR, that it

1   combines real and virtual objects in a real environment
2   runs interactively, and in real time
3   registers (aligns) real and virtual objects with each other.

Azuma and his colleagues (2001) also considered the removing of real objects as a subset of AR. And so, active noise cancelling (ANC) and other sound 'erasing' capabilities have often been thought of as part of AAR.

The requirement for augmentations to happen in real-time is crucial: for example, adding sounds to a recording afterward would not qualify as AAR. It would be merely post-production. This also reflects another requirement for AR and AAR: the virtual content must be designed to be a part of—or have a contextual relationship with—the environment (Naphtali and Rodkin, 2019). Listening to music while walking through a real environment may produce new interpretations of the surroundings or form new meanings inside the listener's head, similar to AAR, but these juxtapositions would merely be unintentional (Cliffe, 2022). Also, sometimes the play-back audio content may momentarily and accidentally match with the real environment both spatially and contextually, like hearing a firetruck siren in an audio drama while walking on a street. However, when the content is designed to align with the user's current state and location, a temporal synchronisation and real-time experience is achieved, even in a linear format like an audio walk.

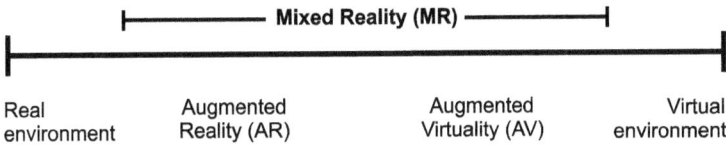

Figure 2.3 A simplified representation of the reality-virtuality continuum (adapted from Milgram et al., 1994)

2.2.2 Mixed reality

Mixed reality is a concept introduced in the 1990s, and, according to the seminal reality-virtuality continuum by Paul Milgram and Fumio Kishino (1994; Milgram et al., 1994), it can be used as the blanket term for any form of augmentations between the real environment and completely virtual environment (virtual reality).

According to the model, if the observed environment is principally real, the application is called augmented reality, and if most of the observed surroundings are virtual, it is called *augmented virtuality* (AV). An example of AV could be the integration of the user's own voice or footsteps in a VR experience (Larsson et al., 2010). At what point something becomes AR, MR, or AV, is not tightly defined, as discussed earlier.

2.2.3 AR and local presence

During recent years, another dual-letter abbreviation, XR, standing for *extended reality*, has gained popularity as an umbrella term to embrace all the above-mentioned paradigms—VR, AR and MR. With new devices, applications, and modes of use of these media, there has also emerged a need to scrutinise old theories and rethink the relationships between all these new concepts. One such attempt has been made by Rauschnabel et al. (2022) who looked critically at the concepts of XR, MR, VR, and AR, and ended up re-organising them into a new framework. During that work, they separated AR and VR as independent continua contrary to Milgram and Kishino, justifying this by the notion that in practice, their design goals and user experiences are rather different. For instance, according to them, an application is either AR or VR, but never both at the same time.

While their definition of AR largely follows that of Azuma and colleagues', they refine AR based on the level of local presence, which they see as the principal measure of AR. Local presence refers to how well virtual content appears to be right 'here' in the real environment

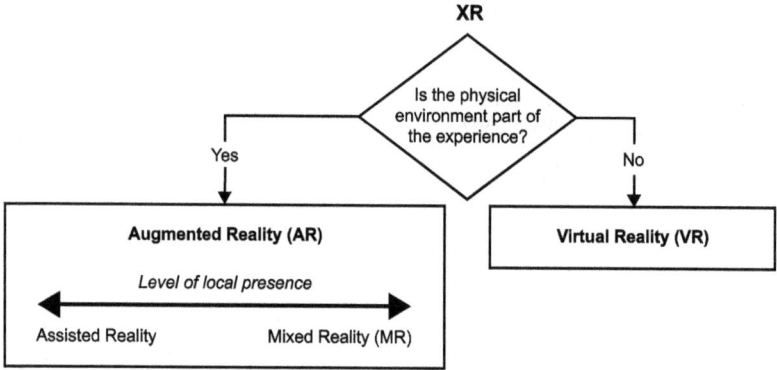

Figure 2.4 XR framework and AR continuum (adapted from Rauschnabel et al., 2022)

(Verhagen et al., 2014; Rauschnabel et al., 2022). In the suggested framework, the scale of local presence starts from *assisted reality* (low) and ends in mixed reality (high).

An example of an auditory 'assisted reality' application would be the tourist audio guide described earlier. While such a system is interactive and the content is rooted in the real-world surroundings, the virtual voice plays *head-locked* in the user's earphones rather than appearing to come from the environment. An auditory MR application, in turn, could be a narrative experience in a museum where the virtual sounds are not just interactive, but emanate from real-world objects with their virtual sound waves propagating through and around obstacles and reflecting from surfaces so realistically that it is almost impossible to distinguish artificial sounds from real ones.

2.3 Application domains

AAR has been used in a number of applications, and it holds a significant potential for wider use. Possible application domains are at least 1) situational awareness, 2) navigation, 3) presentation and display, 4) narratives in education, training, entertainment, art, and social applications, 5) non-narrative art, 6) telepresence, and 7) healthcare (see Yang, Barde and Billinghurst, 2022). Chapters 4 and 5 will present a selection of examples of past and existing applications representative of some of the aforementioned categories.

1   Situational awareness is a domain where AAR fully utilises the omnidirectional nature of hearing; ears can notice events happening

around the user even outside of the field of view and behind obstacles. In some military aircrafts, pilots hear approaching threats as spatialised, three-dimensional sounds, reducing their reaction time significantly (Kucinski, 2018; Evans, 2024). Modern cars are also equipped with in-cabin, audio-based awareness systems, and otherwise quiet electric cars are mandated to produce an artificial sound to warn pedestrians (Thor, 2024). For visually impaired persons, there are applications that describe the surroundings based on video image, or inform about approaching obstacles with sounds (see Akın and Cömert, 2023). Some hearing aids and even consumer-level earbuds are equipped with AI-based technology that isolates speech from background sounds while enhancing it with re-reverberation and other algorithms (Sieber, 2024; Strom and Copithorne, 2024).

2 Satellite navigation apps with voice guidance are among the earliest forms of assisted reality applications, enabling eyes-free operation while navigating to a destination. Whereas existing navigation applications do not yet utilise sound cues that are spatially embedded in the real environment, research has been conducted on that front (e.g., Albrecht, Väänänen and Lokki, 2016).

3 The presentation and display domain refers to applications that use spatialised, environment-embedded virtual sounds to present information that is not necessarily related to the actual surroundings (Yang, Barde and Billinghurst, 2022). An example would be virtual loudspeaker software used in music studios where the engineer can playback multichannel surround mixes through head-tracked headphones without needing the actual, physical loudspeakers in the room (e.g., APL *Virtuoso* or Genelec *Aural ID*). Auditory sticky notes that can be posted on objects in the user's environment would be another example, although such applications remain so far only as research prototypes (e.g., Lokki et al., 2004).

4 Audio augmentation has been used, and can be used, to convey narratives in exhibitions, education and training, attractions, tourist audio guides, sound art installations, research projects, AAR games, audio walks, and social audio sharing platforms. While most applications still lack sophisticated audio spatialisation due to technical bottle necks in user tracking and virtual audio processing, narrative and interactive ideas are developing without the limits of reality. Narratively, one of the most intriguing possibilities is to bring a virtual layer of reality—an auditory one—on top of the existing reality. It can be used to convey forgotten stories that have been suppressed by dominant narratives, or any other perspective to a current matter. With wearable devices, the auditory layers can be

personalised while, at the same time, being unobtrusive to other people. AAR can also foster environmental awareness by enhancing people's perception and understanding of the surroundings (Lawton, Cunningham and Convery, 2020).

5   AAR is also a fruitful platform for non-narrative art. It can implicitly and evocatively comment on peoples' relationship with the environment, or digitally deceive, or just poetically immerse the participants so they observe the real world differently, with new, fused meanings attached to familiar objects.

6   One of the most promising fields for AAR technologies is telecommunications with hologram calls. For modern remote conferencing, to be able to hear other participants as naturally as possible, embedded around the user as if they were there, would help not just intelligibility but also make the interaction and social communication more natural. This dream has been chased since the invention of the telephone in the 1870s, but only now is it becoming reality through common standards and codecs, faster internet connections, and developments in virtual acoustics research.

7   Finally, healthcare is an example of a field where AAR could bring benefits such as real-time auditory feedback for surgeons, cognitive therapy with adaptive spatial audio cues, augmented audiometry tests, tinnitus treatment, and likely many others.

2.4 Location dependency

AAR applications fall roughly into three categories based on how tightly they are based on a certain location. These categories, for their part, have a significant correlation with the chosen technological platform.

1   *Ubiquitous* AAR applications can be used everywhere and all the time, like navigational guides, situational awareness systems, and speech enhancement applications. Such apps would normally run in a user's own mobile and/or wearable device, or be integrated in a vehicle. They would be actively aware of the user's surroundings as well as their location and orientation within the environment. Ubiquitous applications obviously require either a wearable or vehicle-mounted system, which is usually the user's own. Hence, the software and any extensions need to function in multiple devices and technical platforms, respect standards, and be reliable and bug-free. Content-wise, these applications may need to be aware of their surroundings using various sensors and communication with nearby devices. They should also have real-time access to databases

and services such as map, traffic and public transportation data, and cloud-based computing.

2   *Site-specific* experiences are uniquely tailored to a specific location, such as a historical site. These can utilise the user's own mobile device, but they can also be realised with a dedicated system based on either headphones or loudspeaker-based systems, freeing the user from wearing and operating any devices. While site-specific applications can be realised with personal devices, they may as well rely on much more custom solutions, fixed installations, and loudspeakers, although a high throughput of users imposes requirements for solid performance.

3   *Transferable* applications are designed to offer a consistent narrative or experience based on the user's real-world environment, but can nevertheless be moved between different locations with minor adjustments. This category can encompass any platform, depending completely on the application at hand.

2.5 Virtual sounds

Virtual sounds are sounds that are artificially generated or originating from another environment (Härmä et al., 2003) and mediated by some mechanism. A real person howling behind a window in a haunted house experience would not be an AAR experience. However, if the howling were recorded and played back through a loudspeaker behind the window, we could consider that as a virtual sound.

More typically, however, the virtual sounds are generated computationally (see Schraffenberger, 2018; Rauschnabel et al., 2022). This allows more sophisticated real-time modification of the sounds with interactivity that dynamically utilises data from sensors and networks. Also, when using binaural and some other methods such as wave field synthesis, computer technology allows virtual sounds to be spatially aligned with the environment, so that they appear to emanate from the real world and its objects, despite the user's movements in and through space. It may be worth pointing out, that while the sounds are virtual and often imaginary, like the voice of a navigation app, they can still present 'real' information such as directions to a destination (Wellner, 2020).

Terminology-wise, *audio* as a noun refers to recorded, reproduced or transmitted sound. Only when used in combining form, like 'audio perception' or 'audio augmented reality', does it relate to hearing or sound in general. Therefore, the term 'virtual sounds' is used in this book rather than 'virtual audio' to emphasise that while these sounds are artificially generated and mediated by technology, they are intended to blend seamlessly into the environment as if they were naturally occurring *sounds*, thus disguising their nature as mediated *audio*.

In addition to merging, the real soundscape can be modified by filtering out specific sounds from the user's perception, as demonstrated by active noise-cancelling headphones. These removed sounds can then be replaced with virtual ones, if desired. Moreover, if removal is technically unfeasible, deliberate masking can be employed—for example, augmenting the hum of an air conditioner with the sound of waves.

2.6 Plausibility

Plausibility refers to how well a virtual environment aligns with each listener's personal experiences and expectations, rather than aiming for an exact match with reality (Lindau and Weinzierl, 2012). It is therefore a very useful concept for virtual audio researchers and application designers: to evaluate quality of a virtual environment, there is no need to conduct a comparison between the simulation and reality, but merely measure user's subjective perception.

Another concept related to plausibility is *authenticity*, which tests whether the virtual sound is distinguishable at all from the real. One can imagine a situation in AAR where there are two objects side by side, one emitting real sound while the other is augmented with a similar but virtual sound. Now, the listener is able to compare the virtual and the real and decide whether to embrace the illusion or not. However, such cases are likely to be extremely rare in practical AAR applications, which means that authenticity is seldom needed (Meyer-Kahlen, 2024). In most cases, the user is not presented with a real-world comparison, hence they compare the virtual sound with their internal reference, a mental image.

Whereas plausibility relates to virtual representations, *verisimilitude* is a relating concept within the film sound design theory (Chion, 1994); the spectator usually cannot compare the sound on screen with the real one, but refers to their memory of that type of sounds, or conventions established by cinema itself or other cultural products. This notion directs sound designers of all disciplines not to recreate sounds as they are in real life, because the audience may not perceive them as plausible, but instead use sounds that meet the expectations and cultural background of the listener (Lyons, Gandy and Starner, 2000; Stevens, 2009). Film sound tropes demonstrate this well: when a protagonist steps in front of a microphone on a restaurant bandstand and knocks it, we always hear a squeal of feedback. While this rarely happens in real life—or at least should not—that sound of a non-professional PA system is culturally recognisable from community halls and schools, lending the scene an atmosphere of unpolished realism.

A requirement for plausibility and verisimilitude is *congruence* between the elements in question. Congruence (or congruency), in general, can be understood as 'agreement, harmony and appropriateness' (Stevens, 2009, p. 13). In virtual audio research, *audiovisual congruence* often refers to how well the virtual audio source is spatially aligned with the visual object while *room congruence* refers to how well the virtual acoustics match with the real ones. Both of these are important if the illusion of virtual and real coexisting is desired. In Chapter 9.3, *object congruence* is discussed as one of the narrative techniques of AAR, referring to the contextual match between a real-world object and a virtual sound.

2.7 Sonification and sensory augmentation

While many AAR applications and research projects attempt to create an illusion of virtual sounds being real and indistinguishable from the reality, that is often not a desired goal at all. In many applications, a more useful approach is to use *sonifications*. Sonification is usually non-speech audio that translates information or data into sounds that are easy to comprehend by human beings (Nees and Walker, 2011). Sonifications can resemble real sounds and they may be spatialised, but their main purpose, first and foremost, is to inform. The ability to distinguish environment-embedded sonifications from real sounds may be crucial to avoid confusion (Härmä et al., 2003; Dam et al., 2024). An AAR designer should, therefore, be careful when to aim at virtual sounds that are indistinguishable from real ones and when to use sonifications or stylised sound design. This is further discussed in Chapter 8.10.

Basically any data can be sonified: in the 1990s, researchers (Mynatt et al., 1998) at Xerox PARC in California made an AAR prototype that kept office workers informed about the number of emails they had received by playing back sounds of seagull cries. Whereas in their Palo Alto office, the virtual nature sounds functioned as status updates, sonifications are also common as alerts and warnings: when reversing certain car models, the decreasing distance to an obstacle is conveyed through beeps that accelerate as the vehicle gets closer. Sonification can be used to explore data, too; for instance, seismic waves can be translated into sound waves using frequency and time shift, also called *audification* (Nees and Walker, 2011). Sonification is popular in art, too, Ryoji Ikeda's *test pattern* being an example where raw computer data is translated into sounds and music (Rugoff and Bidder, 2023).

Sonifications, or auditory assistance, can help to divide cognitive load from the visual channel to the auditory channel. Many demanding tasks that require situational awareness and sense of spatial orientation, like flying an airplane, are often aided with visual assistance systems. These, consequently lead to high amounts of visual information, potentially creating a bottleneck for cognitive processing (Simpson et al., 2005; Ziemer and Schultheis, 2019). Delivering some of that information through sounds may ease the cognitive load and lead to faster reaction times (Veltman, Oving and Bronkhorst, 2004).

Closely related to sonifications are the concepts of *sensory augmentation* and *sensory substitution*, which can be considered as forms of AR (Schraffenberger, 2018). Sensory augmentation refers to 'super sensing' that enhances human sensory capabilities and enables sensing real but otherwise unobservable phenomena. With 'super hearing', human perception can extend to, for example, ultrasounds by frequency-shifting them down to the audible range (e.g., Pulkki, McCormack and Gonzalez, 2021). A Geiger counter, in turn, allows the user to hear the level of radioactive radiation present at their current location (Schraffenberger, 2018).

Sensory substitution, on the other hand, converts data between two senses. In physiology, it refers to using one sensory modality to compensate for the loss or impairment of another, like using tactile feedback to replace vision (Bach-y-Rita and Kercel, 2003). In AR, however, the term is often used to describe the translation of sensory data into another modality for accessibility or enhancement. One intriguing example of cross-modal sonification involves translating a vehicle's movements—such as an aircraft's attitude, which the human body may not inherently detect—into audible changes in sound or music (Simpson et al., 2005). Assistance applications for people with reduced vision are perhaps more common examples, to be discussed next.

2.8 Accessibility

AAR technologies and applications can offer benefits to people with visual or hearing impairments. Auditory navigational guides and visual interpreters can provide information about the surroundings when vision is unavailable. They use computer vision, data from various sensors, generative artificial intelligence, and map-based information to provide situationally aware sonifications and auditory descriptions (see Akın and Cömert, 2023). Examples of such apps are *Envision AI, SuperSense, Google Lookout*, and Microsoft's *Seeing AI*. Using such apps on a pair of smart glasses frees hands from holding and tapping a mobile phone, thus making everyday life potentially easier. Some apps, like *Obstacle Detector*, use the mobile phone's LiDAR (light detection and ranging) sensor, or 'laser radar', to measure distances to near-by objects and sonify them. For map-based navigation, there are lots of applications, most based on satellite positioning and public map data.

Visually impaired people can also enjoy narrative AAR experiences. As one of the blind participants of *The Reign Union* experience (see Chapter 5.4.) remarked, she had always felt jealous of the augmented reality experiences that she could not take part in—until now!

For people with hearing impairment, technologies associated with AAR such as sound isolation, noise reduction, de-reverberation together with head, eye and location tracking and biosensors can improve speech intelligibility (Georganti et al., 2020) and thus also tremendously enhance quality of life. Some medical hearing instrument manufacturers such as *Sonova* and *Signia* are already equipping their devices with some of the aforementioned technologies utilising real-time artificial intelligence (AI) processing, and similar capabilities are also getting into much lower priced devices such as *over-the-counter (OTC)* hearing aids and even smart earbuds—or *hearables*—tailored with hearing enhancement features (Sieber, 2024; Strom and Copithorne, 2024).

In Chapter 8.9., some design considerations in terms of accessibility are further discussed.

2.9 Speech and dialogue

Human voice and words are central in audiovisual and stage arts, and obviously even more so in audio-based media such as audio dramas, podcasts, audio games, and audio augmented reality. There are logically several reasons for this *verbocentrism* (Chion, 1994). Firstly, spoken language is just very efficient for conveying complex ideas, instructions, and even emotions. Secondly, characters are elemental in storytelling (Green and Appel, 2024), and in audio-based media, they primarily exist through speech and dialogue. The words that characters speak communicate meanings, intentions, and emotions, and they even help the audience to envision what each character looks like (Putt, 2023)—and not just how they look, but also how they smell and feel.

Guidance and other utility applications may not require the use of any spoken information, relying solely on well-designed sonifications that leave little room for misinterpretations. There are also examples of artistically driven experiences that aim at evocative, resonant, and emotional outcomes without trying to explicitly narrate events, such as *The Sound of Things* (Förterer, 2013) and *Walkie-Talkie Dream Angles* (Naphtali and Rodkin, 2019); these will be discussed later in this book.

Even though characters and dialogue are a core element in storytelling, it is definitely possible to create a fully non-verbal story in AAR, based solely on the use of sound effects and music in relationship with the real-world events and environment. While that may be a nice challenge for the creator, the listener may find it demanding to follow such a story, unless it is communicated that free interpretations are 'allowed'.

2.10 Auditory displays

Auditory display, or 'listening interface', is the most significant technical element in an AAR experience, dictating what kind of content can be produced, where, to whom, and how. It is the primary interface through which the virtual sounds get embedded in the user's surroundings. Yet, other components, such as computer, software, and sensors must not be neglected, and they are later discussed in Chapter 7.

Here, the auditory displays are divided into two categories, *binaural* and *soundfield-based* (Larsson et al., 2010). The most typical binaural audio displays are *headphones*, including earbuds and near-ear speakers. These are also the de facto platforms for ubiquitous AAR, and widely used also in site-specific experiences. Binaural audio can also be delivered using *cross-talk cancellation (CTC)*, offering a headphone-free alternative for some special use-cases. Soundfield systems, in turn, encompass a wide variety of loudspeaker-based approaches, of which the most usable for AAR will be discussed in this chapter.

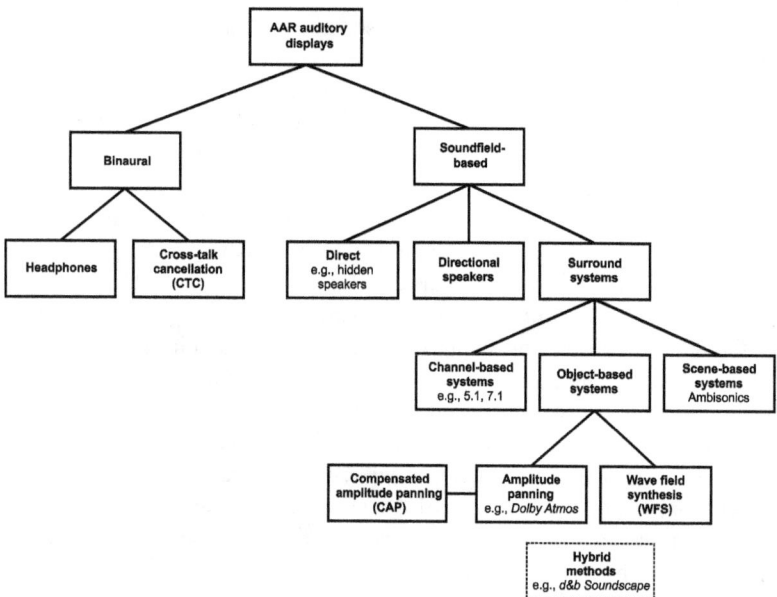

Figure 2.5 Auditory displays and technologies applicable to AAR

2.10.1 *Binaural methods*

'Binaural sound refers to the two-channel sound that enters a listener's left and right ears', and compared to regular stereo, it is filtered in a way that mimics the way head, outer ears, and torso would manipulate sound waves before they enter ear canals (Roginska, 2018, p. 88). These alterations provide cues for the brain to localise sounds. Binaural audio is typically played back directly to the ears using headphones, and it is supposed to sound natural and externalised as if coming from the environment instead of playing inside the user's head.

Binaural audio can be recorded by inserting small microphones in a person's ears, thus creating a natural-sounding replication of what *that* person is hearing including the sense of three-dimensionality. Another way is to use a head-shaped microphone rig called *dummy head*, offering more universally applicable results and avoiding the recordist's own body sounds getting captured. More often in XR applications, however, the audio content is generated and processed computationally from individual virtual sound sources, and the output is decoded into binaural using signal processing. In this rendering phase, *head-related transfer functions* (HRTFs) are used. HRTFs describe how the sound arriving from different angles gets coloured and phased by anatomical features of head, outer ears, and torso. As each person's pinna morphology is unique, the spatialisation gets more accurate when using individualised HRTFs (Sunder, 2022). However, as these personal HRTFs are difficult to measure in practice, generic HRTFs are often used (see Chapter 6.4.).

Headphones

As mentioned, headphones are the most used binaural delivery platform. The term 'headphones' here encompasses all kinds of wearable audio displays, including traditional headphones with a headband, earbuds, open-ear headphones, and near-ear speakers installed in smart glasses such as *Bose Frames* or *Ray-Ban Meta*. Also, neckband speakers resting on shoulders can be discussed together with headphones.

Headphones enable personal and unobtrusive experience, but in MR applications, they also require sophisticated virtual audio processing and head tracking. These and the variety of different headphone types are discussed in more detail in Chapters 6 and 7. One significant parameter of headphones is their acoustic transparency, or how well or selectively the environment can be heard, a key consideration for an AAR designer. This is further discussed in Chapter 3.8.

Cross-talk cancellation

Instead of headphones, binaural audio can be delivered to the user's ears over loudspeakers using *cross-talk cancellation* (CTC, or XTC). If one listens to binaural audio from two loudspeakers, normally, the left speaker sound enters the left ear, but also the right ear. This is called *crosstalk*. For binaural audio, this must not happen, otherwise the spatial effect is lost. The left channel signal to right ear must be *cancelled*, and the same be done for the right channel.

The cancellation can be done with sophisticated filters and computer algorithms. If the listener is allowed to turn the head and move, the head must be tracked while the system attempts to adjust the filter parameters to keep the sounds tightly beamed to the left and right side of the head without spilling to the wrong ear. For instance, the CTC system made by *Audioscenic* tracks the user's head with a camera mounted in between the speaker elements.

The obvious advantage of CTC is that the user is free from any wearable device. However, the current realisations enable only a very limited area of movement for the user before the binaural image collapses, although that is likely to change as the technology advances. Another benefit is that the sound sources natively appear as realistic and externalised, as if coming from outside of the head (Choueiri, 2018). A challenge from the AAR point of view is to conceal the loudspeaker array to maintain the illusion of nonmediation.

2.10.2 Soundfield-based methods

While binaural methods deliver distinct audio signals directly to each ear, soundfield methods use loudspeakers to generate an acoustic field that the user perceives naturally with both ears. Direct augmentation using hidden loudspeakers is the oldest, simplest, and still perhaps the most used method to create an illusion that an inanimate object is 'alive', used in museums and attractions all over the world. Directional speakers are also rather common in public exhibitions. Other soundfield-based formats, however, require installing loudspeakers around the listening area, which may be inconvenient both practically and visually.

Direct augmentation

In direct audio augmentation, the virtual sounds are produced by loudspeakers or other transducers, playing back the audio content *directly* from the real-world object or location. Direct augmentation is a term adapted by the author from Normand, Servières, and Moreau (2012), who use it to refer to augmentation type where information is added to the real world instead of adding it 'between the observer's eye and the real world'. In visually based AR, this is normally achieved by projectors, whilst in AAR, it is achieved by loudspeakers.

Loudspeakers are usually hidden inside or near objects to create the illusion that they are producing the virtual sound. For a plausible illusion, the loudspeaker should be able to reproduce the audio content with high-enough quality. Especially low-frequency content requires large speaker cabinet sizes, which may pose a challenge for concealment. The speaker's directivity pattern can also differ significantly compared to how the real-world object would emanate its sound, which must be taken into account.

If the object has an integrated speaker, that can of course be used, provided that its quality and attributes match with the requirements of the sonic material. Objects can be augmented with multiple speakers, for example, to enable them to sound different from various angles. Loudspeakers can sometimes be replaced by actuators that vibrate a surface such as glass in front of the exhibit (e.g., Devine, 2018). Interactivity is often realised with motion or proximity sensors like in *Barque Sigyn* (see Chapter 4), although nothing prevents using other, more sophisticated logics.

A special case of direct augmentation involves adding sound to a visible loudspeaker. For example, when using a binaural system, adding a virtual sound to a real-world loudspeaker aligns naturally with the concept of AAR. However, playing the same sound directly through a real-world loudspeaker would hardly qualify as AAR, even if it appeared identical to the observer. After all, playing old announcements through a historical PA system in a museum would, in a sense, augment the space with a historical auditory layer.

Direct augmentation offers technical simplicity and robustness, straightforward design, applicability to indoors and outdoors, and requires no wearable devices for users. The fact that everyone within the speakers' reach will hear the same sounds may be seen as an advantage, but also a disadvantage as no individualised content is possible. Loudspeakers in public spaces also pose a risk of cacophony. Further, acousmatic sounds floating in mid-air are impossible to create with this approach (see Chapter 9).

Directional loudspeakers

Directional loudspeakers focus audio to a specific listening area, limiting sound leakage. Most directional speakers send out ultrasound which interacts with air molecules to create audible sound only within a tight cone, utilising the *air nonlinearity* principle (Zhou et al., 2024). Other methods also exist, like using more conventional electrostatic speaker elements (Coppin, 2023). Directional speakers have usually reduced low frequency reproduction, which may be a limiting factor for some AAR applications.

Using the principle of air nonlinearity with two or more directional speakers it is, in theory, possible to create virtual sound spots within a space. Researchers (Zhou et al., 2024) at Shanghai Jiao Tong University have experimented with two phased ultrasound arrays, one transmitting a carrier wave and the other an up-converted audio signal. The waves intersect at a precise point where air nonlinearity generates audible sound, forming a localised sound spot. The system also tracks users in real-time, dynamically adjusting the spot's position. While still at a prototype stage, this approach holds interesting potential for AAR.

Surround systems

Surround systems use loudspeakers arranged around the listening space. There are three main methods to deliver audio through surround systems, *channel-based, object-based,* and *scene-based*. These methods are further discussed in Chapter 6.

In channel-based systems, such as 5.1 and 7.1, the speakers are placed in predefined positions and orientations. Each loudspeaker outputs a distinct audio channel, with the audio content being mixed and panned specifically to these channels. While there are many standardised loud-speaker configurations, custom speaker positions can also be used.

In the object-based audio (OBA) method, individual sounds are freely positioned around the listener, and the systems automatically place these sound objects across the available loudspeakers. One of the first and most influential methods to calculate this has been *vector base amplitude panning* (VBAP) (Pulkki, 1997) with many variations developed since and used in, for instance, *Dolby Atmos, DTS:X,* and *MPEG-H* codecs. Wave field synthesis can also be considered as an OBA system. Some sound reinforcement manufacturers such as d&b, L-Acoustics, Martin Audio, and HOLOPLOT, offer OBA-based systems where live performers are tracked on stage while their amplified voices and sounds are spatially aligned with their positions. Arguably, this approach can be considered as a form of AAR and it has been used to align sounds also with puppets and other objects.

The scene-based approach delivers audio through surround speakers using *Ambisonics,* and particularly the more high-resolution version *higher order Ambisonics* (HOA). Ambisonics is a spatial audio format that is not based on loudspeaker channels but instead uses a mathematical representation of the sound field. This allows flexibility and decoding to any speaker layout or binaural listening.

For most AAR applications, surround systems may not be very practical because, in order to create the illusion of nonmediation, the speaker array should be concealed or their presence justified. Some soundbars attempt to replace some of the surround speakers by utilising wall and ceiling reflections, which may be useful for some applications, but their spatial rendering is not very convincing. Also, most surround systems require the listener to stay within a specific area, the *sweet spot*, to experience the spatial sound scene optimally.

That said, some cars utilise the built-in surround speaker system to play back warning sounds from the appropriate direction. The integrated speakers are also relatively unnoticeable. In the *Audio Nomad* project (Woo et al., 2006), surround speakers were installed on a moving cruising ship deck, spatially aligning virtual audio content on passing-by features, although the speakers were not disguised in any way.

To make the sweet spot larger inside the speaker array, thus allowing the listener to move through the space, *compensated amplitude panning* (CAP) can be tried. This method adaptively adjusts the amplitude levels of each loudspeaker based on the listener's position. The system is typically optimised for a single listener at a time. While CAP holds some potential for audio AR, with the aforementioned limitation of using loudspeakers in general, the author is not aware of it being used for AAR.

An interesting curiosity are wearable surround speakers. Since at least the 1970s, when quadrophonic headphones with two drivers per earcup were introduced, there have been experiments with wearable devices that use multiple speakers to create an immersive soundfield. In the 1990s, the Institut für Rundfunktechnik in Germany did tests with a system comprising of little speakers around the user's head. While the system eliminated the use of HRTFs and reportedly worked well, it was commercially unfeasible (Brandenburg and Sloma, 2024). In 1999, Sennheiser released their rather extravagant 4-channel shoulder-mounted speaker apparatus *Surrounder*, which was also briefly tested in an AAR project by Montan (2002). These kinds of devices, if coupled with user pose tracking, could be used as personal auditory displays in some AAR experiences, but before that, a lot of product development would be needed to make them convenient enough.

Wave field synthesis

Wave field synthesis (WFS) is a spatial audio technique that uses dense loudspeaker arrays to recreate the wavefronts of virtual sound sources, creating 'auditory holograms' that appear to emanate from specific points, even inside the room. The sweet spot is large, covering the whole area reached by the loudspeakers. These features make WFS intriguing in terms of AAR.

However, the method introduces several challenges, one being the large number of loudspeakers required, often hundreds, which may be problematic in terms of budget, available space, and narrative justification—unless the story is set in an acoustic laboratory. Another significant challenge is the method's sensitivity to room acoustics, demanding an acoustically dry environment for optimal effect (Sporer et al., 2018). Partly due to these reasons, WTF has remained largely as a scientific and experimental tool, for instance, used for experimental music.

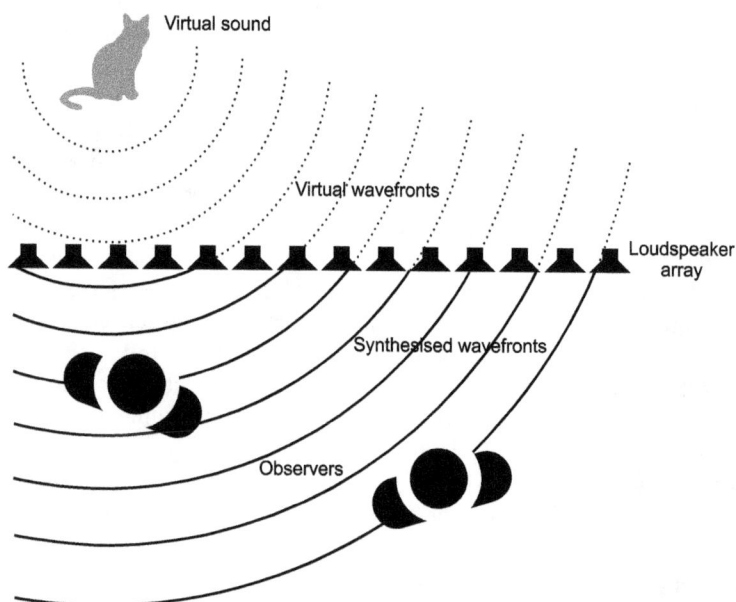

Figure 2.6 An extremely simplified representation of WFS: regardless of the listener's position, the auditory image of the virtual sound remains stable

Recently, however, the concept has been successfully commercialised by HOLOPLOT and Sphere Entertainment, combining the WFS principles with audio beamforming. HOLOPLOT and other WFS-based systems offer many groundbreaking approaches to audio delivery like isolating sounds to tightly defined areas and spatially aligning sounds with objects and people. While isolated sound areas can be achieved with directional speakers, WFS-based systems can be programmed to move them around without changing the placement of loudspeakers.

Hybrid solutions

Some AAR experiences combine different auditory displays. For instance, using acoustically transparent headphones together with loudspeakers allows listeners to enjoy private and individualised content while benefitting from the simplicity and realistic virtual audio reproduction provided by direct, speaker-based augmentations (Mueller-Tomfelde, 2002).

2.11 Externalised and acousmatic sounds

With headphone listening, if no spatialisation is applied to virtual sounds, they often appear as originating from inside the user's head. This *inside-the-head locatedness* (IHL) phenomenon limits the sounds to move only between the left and right ear, also called *lateralisation* (Roginska, 2018). Externalised sounds, on the other hand, appear as originating from the environment, outside of the head. This is naturally achieved by loudspeaker-based systems, but can be realised through headphones as described in Chapter 6.

Further, we can divide the externalised sounds to two categories: *acousmatic sounds* and *attached sounds*. In the context of AAR, acousmatic sounds can be considered as sounds without a physical (and plausible) source (see Cliffe, 2024). Borrowing the term from Pierre Schaeffer, acousmatic sounds traditionally refer to sounds whose cause remains unseen. The acousmatic music practice is characterised by listening to sounds through loudspeakers or headphones without seeing their original sources, allowing listeners to reconstruct the soundscape in their minds (Smalley, 1997).

However, in the context of AAR, if there were visible or tangible loudspeakers in the room, they would form part of the real-world environment. They would be considered as physical objects with distinct sonic properties, rather than remaining a loudspeaker 'veil' to be ignored and merely conveying the acousmatic soundscape. Further, AAR is not just about seeing and hearing, but the real world can be sensed multi-modally:

an object can be unseen but, at the same time, tangible and observable. Therefore, within AAR, we should consider acousmatic sounds in relation to their physical counterparts rather than sound source visibility. Acousmatic sounds, in this context, would refer to those without any physical counterpart in the environment. This perspective accommodates applications set in darkness as well as the ways visually impaired and blind users may experience AAR.

This would also be in line with how the term is used in film studies. There, acousmatic sounds often refer to off-screen sounds, something the camera cannot see (Chion, 1994). However, in AAR, there is no single screen through which the world is watched, but everything around us is the 'screen', everything we can see, hear, smell, taste, and feel. An off-screen sound in AAR would be an 'off-physical' sound, an immaterial sound.

(To avoid confusion with the established use of 'acousmatic', we could as well consider alternative terms for non-physical or non-object sounds in AAR, such as 'holophonic'. While that aptly describes the phenomenon, its adoption has, however, been limited probably because it was once a registered trademark, although now expired.)

Attached sounds, in turn, would be sounds that *do* have a physical counterpart. With direct augmentation, attached sounds are simply realised by attaching a loudspeaker inside or near the object. In binaural systems, accurate spatialisation and externalisation are required to achieve the same effect, to spatially align the virtual sound with its corresponding physical object. Acousmatic and attached sounds in narrative contexts are discussed in detail in Chapter 9.

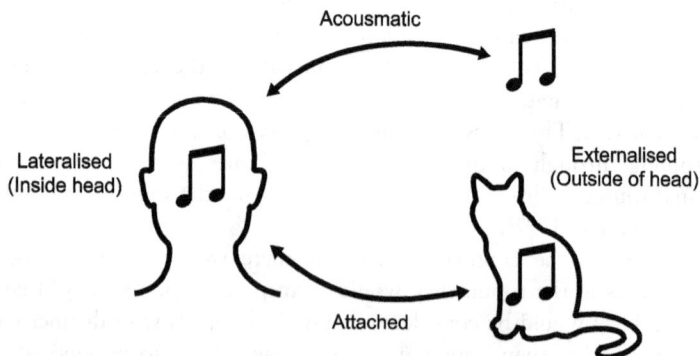

Figure 2.7 Acousmatic and attached sounds as lateralised and externalised

Figure 2.7 presents the relationship between these four concepts, lateralised, externalised, acousmatic and attached sounds. A lateralised sound (playing inside the user's head) can be either acousmatic or attached depending on the context. For example, a voice over is usually considered as acousmatic, or without having a physical origin, although we hear it as coming from inside our head. However, we are usually so accustomed to headphone listening with lateralised sounds and voices that we consider the sound to exist somewhere else, either in the 'aether' or coming from the computer or mobile phone screen thanks to the ventriloquism or 'magnetisation' effect (see Chapter 4). An example of an attached sound playing inside the head, in turn, could be a physiological sound, an internal voice, or perhaps a character inside one's skull.

2.12 Six degrees of freedom

AAR is not just about virtual sounds embedded in the real world; it is also about moving through the environment. Movement enhances the sense of presence (Järvinen, 2017), and with binaural systems, registering movements helps externalisation of virtual sounds, making them appear to really coexist with the reality (Roginska, 2018; Brandenburg, et al., 2023). Whereas the auditory display is the primary *output* interface in AAR, movement is often the primary *input* interface; events and interactions are triggered based on users location and movements.

In the XR world, the concept 'degrees of freedom' is essential. It is an engineering concept referring to different movements that are possible for an object in a given space (Paterson and Llewellyn, 2022). In AAR, where the user—the 'object' in this case—is seldom physically restricted to move in the environment, the degrees of freedom usually refer to the system's ability to track the head of the user and adjust the binaurally reproduced virtual soundscape accordingly.

If the binaural application was designed for a static user who keeps their head straight forward all the time, the virtual soundscape could be spatially fixed, utilising 0 degrees of freedom. If the user remains still, the illusion of virtual sounds coexisting with the reality may sound very plausible, *Séance* (see Chapter 5.4.) being a good example. In a similar application, a virtual sound that was attached to an object 2 metres to the left would appear as coming from that object 2 metres to the left. However, should the user decide to turn their head, the soundscape would turn with it, and the virtual sound would no more be aligned with the object.

To make the soundscape stay fixed even when turning the head, the system should be able to track in *three degrees of freedom* or *3DoF*. The three degrees in this case refer to orientation in Euler angles, or rotational movement: pitch, yaw, and roll. For the system to allow 3DoF, the user's head must be tracked with sensors so that when the head turns, the virtual soundscape counter-rotates; the virtual sound would appear as staying attached to the object.

Now, if the experience is able to track both the user's movement through space and head movements, we have *six degrees of freedom*, or *6DoF* (see Figure 2.8). The three additional degrees are translational movement in Euclidean space along the x axis (left and right), y axis (front and back), and z axis (up and down) (Mazuryk and Gervautz, 1999). Now, the virtual soundscape can be spatially updated to match with the user's movements and it appears as staying fixed relative to the environment.

It is worth noting, that the term 3DoF can sometimes refer to translational movement—movement along the x, y, and z axes—though this usage is rare in the context of virtual audio.

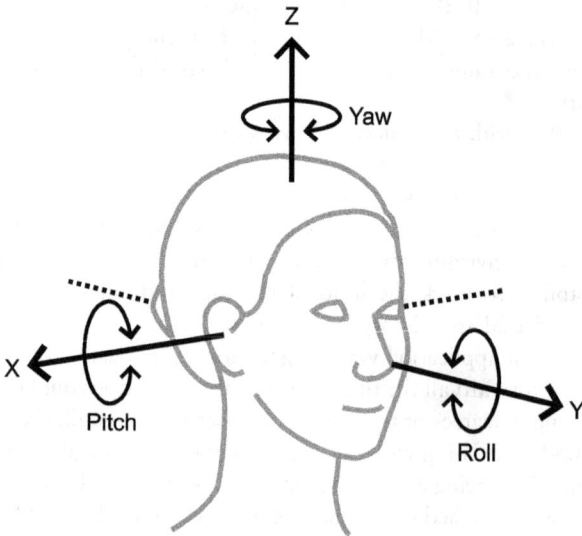

Figure 2.8 Six degrees of freedom (6DoF)

XY-Yaw

Depending on the application, not all degrees need to be tracked. For example, elevation (z) is often omitted. Some tracking systems are not capable of elevation tracking, or by omitting elevation, the translational tracking becomes significantly more accurate. This may be true especially with some radio frequency based tracking systems. Also, if the user is supposed to move on a single elevation plane only (one floor, for instance) and they are not supposed to crouch or climb on ladders, the system can be simplified by omitting elevation tracking.

Similarly, some applications may omit pitch and roll, as they primarily influence the perceived elevation of sound—for instance, a leftward head roll makes left-side sounds seem higher—and, in any case, many binaural systems already face challenges in rendering height differences correctly.

Hence, for many use-cases it is enough if the system allows translational movement along axes on a horizontal plane (x, y) as well as rotation around the vertical axis (yaw). Normand, Servières and Moreau (2012) refer to this as 2D+θ (where the Greek alphabet θ denotes yaw), but in this book, it is described in simpler terms as XY-Yaw.

2.13 Conclusion

AAR can be understood as a framework of multiple technologies and applications with a shared goal: enhancing the perception and understanding of reality. This is accomplished by merging virtual auditory content with the environment, erasing and replacing sounds, and improving hearing. This chapter introduced some of the key concepts and terminology related to AAR, which will be referenced throughout the book. The next chapter explores reality, immersion, and interactivity, establishing the foundation for augmented reality applications and experiences.

References

Aalborg University (2024) 'Audio Only VR for Blind Gamers', Aalborg University's Research Portal. Available at: https://vbn.aau.dk/en/projects/audio-only-vr-for-blind-gamers (Accessed: 3 December 2024).

Akın, A.T. and Cömert, Ç. (2023) 'The Development of an Augmented Reality Audio Application for Visually Impaired Persons', *Multimedia Tools and*

Applications, 82 (11), pp. 17493–17512. Available at: https://doi.org/10.1007/s11042-022-14134-x.

Albrecht, R., Väänänen, R. and Lokki, T. (2016) 'Guided by Music: Pedestrian and Cyclist Navigation with Route and Beacon Guidance', *Personal and Ubiquitous Computing,* 20 (1), pp. 121–145. Available at: https://doi.org/10.1007/s00779-016-0906-z.

Azuma, R. (1997) 'A Survey of Augmented Reality', *Presence: Teleoperators and Virtual Environments,* 6 (4), pp. 355–385. Available at: https://doi.org/10.1162/pres.1997.6.4.355.

Azuma, R. et al. (2001) 'Recent Advances in Augmented Reality', *IEEE Computer Graphics and Applications,* 21 (6), pp. 34–47. Available at: https://doi.org/10.1109/38.963459.

Bach-y-Rita, P. and Kercel, S.W. (2003) 'Sensory Substitution and the Human-Machine Interface', *Trends in Cognitive Sciences,* 7 (12), pp. 541–546. Available at: https://doi.org/10.1016/j.tics.2003.10.013.

Brandenburg, K. et al. (2023) 'Implementation of and Application Scenarios for Plausible Immersive Audio Via Headphones', in *Audio Engineering Society Convention 155,* Audio Engineering Society. Available at: https://www.aes.org/e-lib/browse.cfm?elib=22308 (Accessed: 25 November 2023).

Brandenburg, K. and Sloma, U. (2024) 'Conversation with Matias Harju, May 23'.

Chion, M. (1994) *Audio-Vision: Sound on Screen.* Columbia University Press.

Choueiri, E. (2018) 'Binaural Audio Through Loudspeakers', in A. Roginska and P. Geluso (eds) *Immersive Sound: The Art and Science of Binaural and Multi-Channel Audio.* Routledge, pp. 124–179.

Cliffe, L. (2022) *Audio Augmented Objects and the Audio Augmented Reality Experience.* PhD thesis. University of Nottingham. Available at: https://eprints.nottingham.ac.uk/69795/.

Cliffe, L. (2024) 'Into the Here and Now: Explorations within a New Acoustic Virtual Reality', *Leonardo* [Preprint]. Available at: https://nottingham-repository.worktribe.com/output/42480431/into-the-here-and-now-explorations-within-a-new-acoustic-virtual-reality (Accessed: 2 December 2024).

Coppin, H. (2023) 'Introduction to Directional Sound with Panphonics Electrostatic Speakers', 26 June. Available at: https://panphonics.com/uncategorized/introduction-to-directional-sound-with-panphonics-electrostatic-speakers/ (Accessed: 29 October 2024).

Dam, A. et al. (2024) 'Audio Augmented Reality Using Sonification to Enhance Visual Art Experiences: Lessons Learned', *International Journal of Human-Computer Studies,* 191, p. 103329. Available at: https://doi.org/10.1016/j.ijhcs.2024.103329.

Devine, C. (2018) 'Resonant Bodies – Caroline Devine', September. Available at: https://carolinedevine.co.uk/2018/09/15/resonant-bodies/ (Accessed: 16 September 2024).

Evans, S. (2024) 'PTC, BAE Team to Create Augmented Reality Fighter Pilot Helmet'. Available at: https://www.iotworldtoday.com/metaverse/ptc-bae-team-to-create-augmented-reality-fighter-pilot-helmet- (Accessed: 10 October 2024).

Förterer, H. (2013) 'The Sound of Things'. Available at: https://www.foerterer.com/sound-of-things.html (Accessed: 3 December 2024).

Georganti, E. et al. (2020) 'Intelligent Hearing Instruments—Trends and Challenges', in J. Blauert and J. Braasch (eds) *The Technology of Binaural Understanding*. Cham: Springer International Publishing (Modern Acoustics and Signal Processing), pp. 733–761. Available at: https://doi.org/10.1007/978-3-030-00386-9.

Green, M.C. and Appel, M. (2024) 'Narrative Transportation: How Stories Shape How we See Ourselves and the World', in B. Gawronski (ed.) *Advances in Experimental Social Psychology*. Academic Press, pp. 1–82. Available at: https://doi.org/10.1016/bs.aesp.2024.03.002.

Härmä, A. et al. (2003) 'Techniques and Applications of Wearable Augmented Reality Audio', in *Audio Engineering Society Convention* 114, Audio Engineering Society. Available at: http://www.aes.org/e-lib/browse.cfm?elib=12495.

Jacuzzi, G. (2018) '"Augmented Audio": An Overview of the Unique Tools and Features Required for Creating AR Audio Experiences', in *International Conference on Audio for Virtual and Augmented Reality*. Redmond, WA, USA: Audio Engineering Society, p. 8.

Järvinen, A. (2017) 'Design for Presence in VR, Part 2: Towards An Applied Model, Medium'. Available at: https://virtualrealitypop.com/designing-for-presence-in-vr-part-2-towards-an-applied-model-2784bf16a01 (Accessed: 29 November 2024).

Krzyzaniak, M., Frohlich, D. and Jackson, P.J.B. (2019) 'Six Types of Audio that DEFY Reality! A Taxonomy of Audio Augmented Reality with Examples', in *Proceedings of the 14th International Audio Mostly Conference: A Journey in Sound*. New York, NY, USA: Association for Computing Machinery (AM'19), pp. 160–167. Available at: https://doi.org/10.1145/3356590.3356615.

Kucinski, W. (2018) 'A-10C Pilots Will Get 3D-Audio to Increase Situational Awareness'. Available at: https://www.sae.org/site/news/2018/11/a-10c-pilots-will-get-3d-audio-to-increase-situational-awareness (Accessed: 10 October 2024).

Larsson, P. et al. (2010) 'Auditory-Induced Presence in Mixed Reality Environments and Related Technology', in E. Dubois, P. Gray, and L. Nigay (eds) *The Engineering of Mixed Reality Systems*. London: Springer (Human-Computer Interaction Series), pp. 143–163. Available at: https://doi.org/10.1007/978-1-84882-733-2_8.

Lawton, M., Cunningham, S. and Convery, I. (2020) 'Nature Soundscapes: An Audio Augmented Reality Experience', in *Proceedings of the 15th International Audio Mostly Conference (AM'20). AM'20: Audio Mostly 2020*, Graz Austria: ACM, pp. 85–92. Available at: https://doi.org/10.1145/3411109.3411142.

Lindau, A. and Weinzierl, S. (2012) 'Assessing the Plausibility of Virtual Acoustic Environments', *Acta Acustica united with Acustica*, 98 (5), pp. 804–810. Available at: https://doi.org/10.3813/AAA.918562.

Lokki, T.et al. (2004) 'Application Scenarios of Wearable and Mobile Augmented Reality Audio', in *Audio Engineering Society Convention* 116, Audio Engineering Society, p. 9.

Lyons, K., Gandy, M. and Starner, T. (2000) 'Guided by Voices: An Audio Augmented Reality System', in *Proceedings of the International Conference on Auditory Display April, 2000*. Georgia Institute of Technology, Atlanta, Georgia, USA: International Community for Auditory Display. Available at: https://smartech.gatech.edu/handle/1853/50672 (Accessed: 5 November 2018).

Mariette, N. (2013) 'Human Factors Research in Audio Augmented Reality', in W. Huang, L. Alem, and M.A. Livingston (eds) *Human Factors in Augmented Reality Environments*. New York, NY: Springer, pp. 11–32. Available at: http s://doi.org/10.1007/978-1-4614-4205-9_2.

Mazuryk, T. and Gervautz, M. (1999) 'Virtual Reality – History, Applications, Technology and Future', *ResearchGate* [Preprint]. Available at: https://www.researchgate.net/publication/2617390_Virtual_Reality_-_History_Applica tions_Technology_and_Future (Accessed: 10 March 2019).

Meyer-Kahlen, N. (2024) *Transfer-Plausible Acoustics for Augmented Reality*. PhD thesis. Aalto University. Available at: https://urn.fi/URN:ISBN: 978-952-64-1913-8.

Milgram, P. and Kishino, F. (1994) 'A Taxonomy of Mixed Reality Visual Displays', *IEICE Transactions on Information and Systems*, E77-D(12), pp. 1321–1329.

Milgram, P. et al. (1994) 'Augmented Reality: A Class of Displays on the Reality-Virtuality Continuum', *Telemanipulator and Telepresence Technologies*, 2351. Available at: https://doi.org/10.1117/12.197321.

Montan, N. (2002) *An Audio Augmented Reality System*. Master's thesis, Royal Institute of Technology, Stockholm.

Mueller-Tomfelde, C. (2002) 'Hybrid Sound Reproduction in Audio-Augmented Reality', in *Audio Engineering Society Conference: 22nd International Conference: Virtual, Synthetic, and Entertainment Audio*, Audio Engineering Society. Available at: https://www.aes.org/e-lib/browse.cfm?elib=11134 (Accessed: 8 September 2023).

Mynatt, E.D. et al. (1998) 'Designing Audio Aura', in *Proceedings of the SIGCHI Conference on Human Factors in Computing Systems – CHI '98. SIGCHI*, Los Angeles, California, United States: ACM Press, pp. 566–573. Available at: http s://doi.org/10.1145/274644.274720.

Naphtali, D. and Rodkin, R. (2019) 'Audio Augmented Reality for Interactive Soundwalks, Sound Art and Music Delivery', in *Foundations in Sound Design for Interactive Media*. Routledge.

Nees, M. and Walker, B. (2011) 'Theory of Sonification', in T. Hermann, A. Hunt, and J.G. Neuhoff (eds) *The Sonification Handbook*. Berlin, Germany: Logos Publishing House, pp. 9–39.

Normand, J.-M., Servières, M. and Moreau, G. (2012) 'A New Typology of Augmented Reality Applications', in *Proceedings of the 3rd Augmented Human International Conference*. New York, NY, USA: ACM (AH '12), pp. 1–18. Available at: https://doi.org/10.1145/2160125.2160143.

Paterson, J. and Llewellyn, G. (2022) 'Towards 6DOF', in J. Paterson and H. Lee (eds) *3D Audio*. New York, NY: Routledge (Perspectives on music production), pp. 43–63.

Pulkki, V. (1997) 'Virtual Sound Source Positioning Using Vector Base Amplitude Panning', *Journal of the Audio Engineering Society*, 45 (6), pp. 456–466.

Pulkki, V., McCormack, L. and Gonzalez, R. (2021) 'Superhuman Spatial Hearing Technology for Ultrasonic Frequencies', *Scientific Reports*, 11 (1), p. 11608. Available at: https://doi.org/10.1038/s41598-021-90829-9.

Putt, B.M. (2023) *Stories Told Through Sound: The Craft of Writing Audio Dramas for Podcasts, Streaming, and Radio*. Essex, Connecticut: Applause.

Rauschnabel, P.A.et al. (2022) 'What is XR? Towards a Framework for Augmented and Virtual Reality', *Computers in Human Behavior*, 133, p. 107289. Available at: https://doi.org/10.1016/j.chb.2022.107289.

Roginska, A. (2018) 'Binaural Audio Through Headphones', in A. Roginska and P. Geluso (eds) *Immersive Sound: The Art and Science of Binaural and Multi-Channel Audio*. Routledge.

Rugoff, R. and Bidder, S. (2023) 'Ryoji Ikeda Interview: "For me, there's no separation between sound and visuals"', *Fact Magazine*. Available at: https://www.factmag.com/2023/04/13/ryoji-ikeda-interview/ (Accessed: 16 October 2024).

Schmalstieg, D. and Hollerer, T. (2016) *Augmented Reality: Principles and Practice*. Addison-Wesley Professional.

Schraffenberger, H. (2018) *Arguably Augmented Reality: Relationships Between the Virtual and the Real*. PhD thesis. Universiteit Leiden.

Sieber, T. (2024) 'Hearables and Hearing Wearables Technology Guide, Hearing Tracker'. Available at: https://www.hearingtracker.com/hearables (Accessed: 13 October 2024).

Simpson, B. et al. (2005) 'Spatial Audio Displays for Improving Safety and Enhancing Situation Awareness in General Aviation Environments', in *New Directions for Improving Audio Effectiveness. Meeting Proceedings RTO-MP-HFM-123, Paper 26*, Neuilly-sur-Seine, France, pp. 26. 1–26. 16. Available at: https://apps.dtic.mil/sti/citations/ADA454658 (Accessed: 21 September 2024).

Smalley, D. (1997) 'Spectromorphology: Explaining Sound-Shapes', *Organised Sound*, 2 (2), pp. 107–126. Available at: https://doi.org/10.1017/S1355771897009059.

Sporer, T. et al. (2018) 'Perception of Spatial Sound', in A. Roginska and P. Geluso (eds) *Immersive Sound: The Art and Science of Binaural and Multi-Channel Audio*. Routledge, pp. 311–332.

Stevens, M. (2009) *Music and Image in Concert: Using Images in the Instrumental Music Concert*. Music and Media, Sydney.

Strom, K. and Copithorne, D. (2024) 'Best OTC Hearing Aids of 2024: Price and Sound Comparisons, Hearing Tracker'. Available at: https://www.hearingtracker.com/otc-hearing-aids (Accessed: 13 October 2024).

Sunder, K. (2022) 'Binaural Audio Engineering', in J. Paterson and H. Lee (eds) *3D Audio*. Abingdon, Oxon; New York, NY: Routledge (Perspectives on music production), pp. 130–159.

Thor (2024) 'THOR AVAS – Acoustic Vehicle Alerting System for All Types of Electric Vehicles'. Available at: https://thor-avas.com/legislation/regulation-on-avas-requirements/ (Accessed: 25 September 2024).

Tikander, M. (2009) 'Usability Issues in Listening to Natural Sounds with an Augmented Reality Audio Headset', *Journal of the Audio Engineering Society*, 57 (6), pp. 430–441.

Veltman, J.A., Oving, A.B. and Bronkhorst, A.W. (2004) '3-D Audio in the Fighter Cockpit Improves Task Performance', *The International Journal of Aviation Psychology*, 14 (3), pp. 239–256.

Verhagen, T. et al. (2014) 'Present it Like It Is Here: Creating Local Presence to Improve Online Product Experiences', *Computers in Human Behavior*, 39, pp. 270–280. Available at: https://doi.org/10.1016/j.chb.2014.07.036.

Wellner, G. (2020) 'Postphenomenology of Augmented Reality', in H. Wiltse (ed.) *Relating to Things: Design, Technology and the Artifact.* Bloomsbury Visual Arts, pp. 173–187.

Woo, D.et al. (2006) 'Audio Nomad: Institute of Navigation', *Proceedings of the Institute of Navigation – 19th International Technical Meeting of the Satellite Division, ION GNSS 2006*, 5, pp. 3117–3123.

Yang, J., Barde, A. and Billinghurst, M. (2022) 'Audio Augmented Reality: A Systematic Review of Technologies, Applications, and Future Research Directions', *Journal of the Audio Engineering Society*, 70 (10), pp. 788–809.

Zhou, J.et al. (2024) 'Visar: Projecting Virtual Sound Spots for Acoustic Augmented Reality Using Air Nonlinearity', *Proceedings of the ACM on Interactive Mobile, Wearable and Ubiquitous Technologies*, 8(3), pp. 1–147. Available at: https://doi.org/10.1145/3678546.

Ziemer, T. and Schultheis, H. (2019) 'Psychoacoustic Auditory Display for Navigation: An Auditory Assistance System for Spatial Orientation Tasks', *Journal on Multimodal User Interfaces*, 13 (3). Available at: https://doi.org/10.1007/s12193-018-0282-2.

3 Reality, presence and interactivity

The most important element of any AAR application is reality. While the connection to reality naturally develops in many applications, such as ubiquitous guidance and situational awareness apps, in narrative experiences there is a risk that reality becomes a mere backdrop, turning the application into a form of audio VR. This chapter will explore reality, the sense of presence, and listening to the environment, and their relationship to AAR.

3.1 Reality

Reality forms the essential foundation of augmented reality. In most cases, AAR attempts to manipulate the user's perception of the three-dimensional world around them: physical objects are augmented with virtual sounds, unwanted real-world sounds are erased, and unobservable events—real or fictional—are revealed by sonification. As part of the environment, AAR can also augment other people as exampled in *Séance* and *The Reign Union* (see Chapters 5.4 and 9.3). Besides external world, the user's own body, clothes and immediate items can be augmented with virtual sounds, such as footsteps, bodily sounds, thoughts.

According to Maurice Merleau-Ponty (2013), reality is not a detached, objective entity that exists independently of human perception. Instead, it emerges through our embodied engagement with the world, shaped and interpreted within the context of our prior experiences. When our senses are augmented with virtual content and interactions, these augmentations arguably become part of our lived reality. From this phenomenological perspective, the physical world exists, but it is always mediated by our sensory and bodily engagement.

These augmentations seamlessly integrate into our pre-reflective lived reality, as perception within this framework does not distinguish between the 'natural' and 'virtual', but instead actively constructs a

DOI: 10.4324/9781003627289-3

coherent experience of the world. In practice, however, if a clear distinction between the 'real' real and the virtual is necessary in AAR, techniques such as sonifications or stylistic content can be used, as discussed in Chapter 2, to force cognitive separation. Similarly, interface sounds can be played back head-locked to differentiate them from spatially rendered virtual as well as real-world sounds.

3.1.1 Other observable environments

The reality does not always have to be the tangible, spatial world around and within us; the augmentation can superimpose content in another observable environment. For example, many live sports telecasts and studio-based television shows are overlayed with digitally created three-dimensional graphics that appear to coexist with the real persons (e.g., Azuma et al., 2001; Schmalstieg and Hollerer, 2016). Instead of augmenting the user's own environment, they augment the televised environment, observed through the screen.

Analogous to this, we can augment a purely auditory environment such as the audio feed of a telephone conversation: the caller can create the illusion of being in a different location by holding a portable audio player and playing back sounds from another place while speaking on the phone. This tricks the person on the other end into believing the

Figure 3.1 Three-dimensional AR graphics merged with live people in a television show
Source: Screen capture provided by the Finnish Broadcasting Company (Yle), with 3D graphics by Julia Tavast

caller is in the location suggested by the audio, rather than where they actually are. While this telephone example, at first glance, seems to lack the three-dimensionality of the television augmentations, we must first remember that the television screen is not three-dimensional either, but two-dimensional and monoscopic; the illusion of spatiality is created by visual cues, enhanced by camera movements and occlusion effects that the digital objects accurately obey. Somewhat similarly, the monophonic sound entails various spatial cues, including the loudness of the sound as well as the amount of room reflections and reverberation (Blauert, 1997).

Augmenting telephone feed is, of course, something of an oddity in the context of AAR, yet it warrants recognition since it is analogous to the television examples and creates an illusion that is spatially aligned with a real environment. The *Zombies, Run!* running game (see Chapter 5) and particularly its *radio mode* uses a similar concept where a personal headphone-listening experience is augmented. The game immerses the user in a zombie-infested world, created solely through audio, and gives the user missions where they have to run and avoid the chasing undead. Between the missions, short 'radio broadcasts' are inserted between the songs played from the user's own playlist. The hosts, two fellow survivors, keep up the spirits and inform about recent zombie activity. Every now and then, the show hosts comment on the music: 'That was a classic' or 'Next one is for everyone out there' as if they had chosen the songs in the playlist.

In this example, the audio augmentation is not superimposed onto the real environment *around* the user, but rather onto the user's personal auditory space, namely the 'headphone reality'. This reality gets augmented by the prerecorded radio broadcast sequences, and consequently—should the user willingly suspend their disbelief—the playlist transforms as a part of the story world. The fact that the show hosts share the music taste with the user feels like a happy coincidence. The physical environment, the streets, forests or landscapes the user is running through, do all still have an important role: as the radio hosts address the user with a line such as 'The next song is for all our runners out there' and reminds them to 'be careful', the environment is highlighted as the potential place where zombies may attack and start chasing at any given moment.

3.1.2 Level of realness

While Milgram and Kishino (1994) discuss about *reproduction fidelity*, or how well the AR display can reproduce the 'images' of the real-world objects, the 'realness' of the reality can also have different levels. As an example, using AAR, we can add sounds to a real-world scene where two persons are having an animate discussion behind a window. As their voices are inaudible, our AAR application can replace the discussion with a virtual one. The reality remains as real as it gets, just the voices are fake.

Now, let us have the same scene acted out by two actors. While it appears the same, the event is not as 'real' anymore. While the actors are still real humans, bringing a level of 'liveness' with their potential connection to the participant (Auslander, 2022), their performance is still a representation and interpretation of real characters and events.

Next, instead of actors, we can use two humanoid robots, or animatronics, mechanical puppets familiar from amusement parks. The characters would be physical and tangible, able to mimic human actions, but lack genuine interaction and emotional expression. Further down the path, we can replace the animatronics with a video projection, or a hologram. Now we loose even the tangibility. Yet, in all of the versions we can still apply the original sound design and utilise the same characteristic concepts of AAR.

This continuum can also be observed with environments. Ubiquitous AAR operates in the real, every-day environment, as do many site-specific narrative experiences. The environment can, however, also be entirely fabricated, such as a stage set or an exhibition installation. Even so, these surroundings are still regarded as real, albeit perhaps to a lesser extent.

However, as Auslander (2022) points out, what constitutes 'real' is influenced by cultural contexts and conventions. As technology advances, forms like video projections or XR experiences can evoke feelings of presence and engagement, challenging traditional notions of liveness, authenticity, and realness. This 'reskinning' or representation of reality is the essence of immersive theatre and many amusement park attractions, museum exhibitions, art installations, escape rooms, and a wide number of other narratively driven experiences.

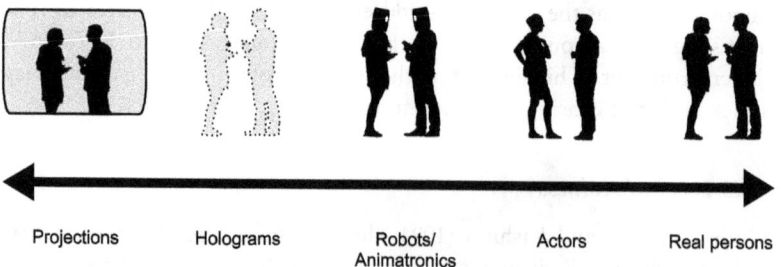

| Projections | Holograms | Robots/ Animatronics | Actors | Real persons |

Figure 3.2 Same event enacted with varying levels of 'realness'
Source: Created by author with human figures designed by rawpixel.com/Freepik

3.2 Sense of presence

A key parameter in AAR, especially when moving towards MR, is sense of presence, often used interchangeably with 'immersion'. Immersion and presence can, however, be considered as separate concepts: immersion referring to the system's technical ability to create a virtual environment, and presence being the human reaction to that, the sense of 'Wow, it's just like being there' (Slater, 2003). This book attempts to follow that distinction, although it may sometimes unintentionally blur the line especially when moving out of the XR context.

3.2.1 Illusion of nonmediation

Lombard and Ditton (1997) presented a widely used definition for presence as 'the perceptual illusion of nonmediation'. In other words, the illusion happens when the person becomes unaware of the mediation system between them and the virtual world (Larsson et al., 2010). An example of when the illusion may collapse is when the user detects loudspeakers in the environment, particularly at the direction of the virtual sounds. It may become challenging to convince the user that the sounds are *not* emanating from the loudspeakers—especially if they actually are. The same may be true if the user has to wear headphones which constantly remind them of the mediation in process.

The sense of presence can be understood as getting formed through a two-level process: 1) the user perceives the mediated environment as a plausible space, and 2) the user experiences themselves as being located within that perceived space (Wirth et al., 2007; Cummings and Bailenson, 2016). Sounds play an important role in this: even though people are usually highly visually oriented, it is suggested that hearing has a crucial role in forming a full sense of presence (Larsson et al., 2010; McGill et al., 2020). First, while sight has a limited field of view, the auditory system is omnidirectional and provides information from everywhere; it can both locate objects and feel the space through reflections and other acoustic cues. Second, while objects in the environment may be visually completely static, the sounds they produce or reflect are constantly ongoing, reminding us that they are there.

For the user of AAR to be able to enter level 1, the sonic augmentations and their spatialisation within the real environment must be plausible enough. They must have the feeling of being here (Lombard and Ditton, 1997) with a level of local presence (Verhagen et al., 2014). Although the auditory system plays a crucial role in creating a sense of presence, vision typically dominates human perception. As a result, even if the sounds are incongruent with the visually perceived world, the sense of presence is likely to be preserved, even with less authentic virtual audio (Stevens, 2009). With blind individuals or if the AAR experience takes place in darkness, the role of sounds becomes enhanced, consequently increasing the requirements of virtual sound reproduction.

3.2.2 Living environment

The sense of being inside the accepted environment, level 2, works through interaction and affordances (Wirth et al., 2007; Cummings and Bailenson, 2016). Affordances are possibilities to act, offered by an object or environment. One important affordance the environment offers is the ability to move through the space and physically interact with the surroundings and objects (Järvinen, 2017). By enhancing the environment and objects with sonic augmentations, AAR creates new affordances. For instance, while a standard office wall might typically go unnoticed, the user might hear muffled sounds of violence coming from the other side—even if they are purely virtual. This auditory cue could provoke a range of immediate responses from the user, such as rushing to the wall to press their ear against it for a clearer listen, searching for an entrance to the room behind the wall to intervene, calling the police, or even choosing to remove their headphones. The augmentations, hence, have a huge potential to shape our actions and interactions within the environment (Turner, 2022).

Besides affordances, the overall liveliness of the environment with things happening beyond the user's control enhances the plausibility illusion (Cummings and Bailenson, 2016). This would, in general, be obsolete in augmented reality as there is no need to make the reality any more real than it already is. However, if the AR system attenuates stimuli from the real world—like an AAR application with acoustically isolated headphones—recreating the liveliness may be beneficial.

There are also other mechanisms that may enhance the sense of presence. For example, intellectual and emotional connections to the experienced world are meaningful (Järvinen, 2017), such as getting into the state of *flow*. In a flow state, the person becomes one with what they do, when they work on a challenge that is more or less in balance with their abilities (Csikszentmihalyi, 1990).

In narrative AAR applications, the most important factor for sense of presence is the story. Narrative immersion, or *narrative transportation*, can happen through various routes, perhaps thanks to a well spatialised audio and strong production values, but more likely through the coherence of the story, character development, emotional intensity, and various other features of a high quality story (Green, 2021).

3.2.3 Suspension of disbelief

Narrative immersion makes us engage in the story even though we know the characters are actually actors, and their voices are recordings played back through headphones or loudspeakers. The spatialisation may not even be perfect, but we *willingly* suspend our disbelief, use imagination, and 'play along' with the story.

With good quality virtual audio rendering and thought-out content design, it is quite easy to use AAR for plausible 'magical' illusions. Magic, however, cannot rely on willing suspension of disbelief: the trick must be so convincing and narratively packaged that even the most sceptical audience member will *unwillingly* suspend their disbelief and start believing in the impossible (Kuhn, 2019). Yet, as discussed in Chapter 2, when there is no real-life reference to compare with, the virtual version of a sound is more easily considered as plausible.

However, as Kuhn (2019) points out, magic depends on what people believe to be impossible, not what is actually impossible. With AAR, if a listener is not aware of the possibilities of virtual audio technology, they may think that it is impossible to make a cat bark and will feel amazed when that happens. On the other hand, the sound of footsteps from the apartment above will elicit wonder only if the listener knows that there is no one at home.

3.3 Interactivity

Interactivity is usually considered as an integral part of AAR (e.g., Yang, Barde and Billinghurst, 2022; Dam et al., 2024). It makes the application useful and integrates it into the user's life. Also, as discussed, the possibility to interact within the environment reinforces a sense of presence.

3.3.1 Two-way process

Interaction is a two-way process where the system takes an input, processes it, and produces an output, or in other words, is able to 'listen, think, and speak' (Crawford, 2012). An automated museum audio guide is interactive; it listens to the positional data sent by proximity or location sensors, decides whether the visitor is close enough to an exhibit, and finally speaks out the voice description to the user's headphones. If the visitor likes what they are hearing, they keep stationed and listening; if not, they walk away, thus making the narration stop. Traditional films, paintings and books also manage to evoke reactions in audience and spark discussion, but they are not interactive because they are just 'speaking' without having conversation; the pieces do not participate in the interaction in any way (Crawford, 2012).

However, interaction requires active participation, and people are different in terms of how much they enjoy using cognitive effort. This is especially true with interactive narratives that demand thinking and engagement. Individuals with a high need for cognition are likely to enjoy them more than others, and thus, even the best interactive story may not work for everyone (Green, 2021). One can argue that in some site-specific narrative experiences, a strong sense of presence can be created without any interaction between the user and the system. *Séance* by DARKFIELD (see Chapter 5) is an example of a linear experience with head-locked, binaural sound that manages to evoke a strong feeling of embodied presence through other methods than direct interaction, such as plausible audio rendering, acting, story, and awareness of the other patrons.

Interaction can also be literally conversational, which is now easily achieved using large language models and other AI tools. While that approach can feel natural and encourage playful engagement, especially in guidance systems (e.g., Wakkary et al., 2004), there is a risk of distraction due to excessive irrelevant information (Breunig, 2024).

3.3.2 Interactional modalities

Interaction with an AAR application can be practiced through several ways and modalities. Some typical examples are discussed next, while technical sensors are explored further in Chapter 7.2.

Geolocated and other location-aware AAR systems know the user's position in relation to the environment and sounds—virtual and real—within it. In many locative applications, walking is the key modality to move within the space (see Cliffe, 2024). Combining kinaesthetic movement with listening to virtual sounds exemplifies cross-modal interaction, where the perception of the overall experience differs from that of its individual components in isolation (Schraffenberger, 2018). When designing walk-based experiences, the user's pace and time spent in specific locations must be well taken into account since sound is temporal in nature and cannot easily be slowed down, accelerated, or paused based on the user's speed. Attractor sounds, as discussed in Chapter 9.12, are one way to approach this challenge.

Besides walking, interaction can be based on any other modality, such as running, bicycling, driving, and cruising, to mention a few (examples of all in Chapter 5). In some cases, like *Zombies, Run!*, only the user's pace is utilised for interaction, omitting positional data.

Some systems can read hand gestures, body posture, and head orientation. Typical examples are pinch and other finger gestures, supported by MR headsets and smart glasses, as well as head nods and shakes, recognised by Apple's *AirPods Pro* earbuds and some other devices. Head orientation can do more than just rotate the soundscape in a dynamic binaural audio display; it can also enable various narrative interactions. For instance, when another character is facing away, a virtual vocalisation can be linked to them (see Chapter 9.3.).

User's voice has become an important interaction modality in AAR with advances in speech recognition. Besides voice commands such as 'Hey Meta, volume up' on *Ray-Ban Meta* smart glasses, voice can also be used more creatively: Spoken words may trigger plot points and be replayed later as part of the narrative. The tone and amplitude of the user's voice—and any other sound, for that matter—could dynamically alter the virtual soundscape by adding effects. This concept is reminiscent of the *RjDj*

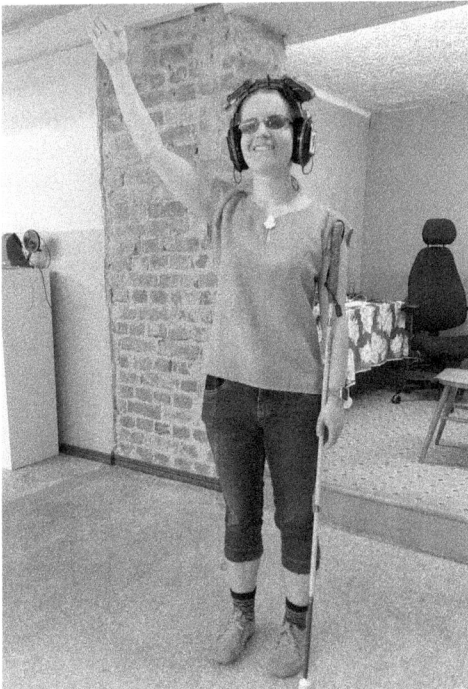

Figure 3.3 Participant's hand gestures in *The Reign Union* are detected and used for interaction with the story
Source: Photograph by the author

mobile app from 2009, which used the phone's microphone and accelerometer to manipulate electronic music based on the user's sonic environment and movements (Cutler, 2009).

Finally, wearable biodata sensors, such as watches and rings, open intriguing possibilities for integration into ubiquitous AAR systems. For example, a user's emotional state can be detected from smart watch data, including measurements of skin temperature, electrodermal activity, heart rate, and heart rate variability (Siirtola et al., 2023). This emotional state could then dynamically influence the audio content, alter the tone of spoken directions, or modify virtual characters' dialogue lines, to name a few possibilities.

3.4 Interactive storytelling

Interactivity in narrative contexts often refers to the participant's ability to manipulate the storyline through their actions and reactions (Crawford, 2012; Glassner, 2017; McErlean, 2018). Interactive experiences can be categorised according to their level of reactivity (Glassner, 2017): In static, 'one-way' experiences such as novels, the information flows in one direction without the possibility of changing it. Choose-your-own-adventure books give the reader narrative choices within tight limits. In 'rail ride' narratives the sequence of events is more or less fixed, even though the user may have freedom to experience them at their own speed. 'Riding the current' offers more possibilities to steer and even fight the flow for some time, but the narrative current is designed to eventually lead everyone along the same path. Many exhibitions with a defined direction of movement are examples of this. The 'two-way' experiences on the other end of the continuum, such as participatory theatre and role-playing games, require a lot of engagement, conversation, and/or exploration.

3.4.1 Demanding art

As mentioned, interactive experiences are demanding to the participants (Glassner, 2017; Green, 2021). That may be the reason why one-way experiences such as books, films, and podcasts are mainstream, whereas role-playing games and other two-way art forms and media 'are still picking up speed' (Glassner, 2017, p. 25).

Participatory, interactive stories also place high demands on their creators (Crawford, 2012). The difficulty of the art is likely another reason for the slow take-off of the truly two-way media. It can be argued that many 'interactive' mediums actually provide quite limited possibilities for users to interact with or manipulate the story itself, but rather offer ways to customise how the experience looks or sounds within a tightly set framework. Glassner (2017) points out that 'The problem with giving the audience control of the story is that they will often, quite reasonably, act to reduce tension and avoid conflict. But tension and conflict are at the heart of great stories' (p. 18).

In terms of user experience, offering an *illusion of agency* can sometimes suffice. Fendt et al. (2012) found no difference in participants' sense of agency when comparing a truly branching text-based game with a non-branching version that provided immediate feedback but did not alter the story's outcome. This approach reduces the authorial burden, as most content typically remains unexplored.

3.4.2 6DoF and storyworld

Even though there is limited possibility of affecting the storyline itself, the participant can usually interact with other elements. Six degrees of freedom (6DoF) offers a natural way to interact with the world: through exploring and changing perspectives the user can discover new information, which can change the way characters, subjects or the story are interpreted. Consequently, this enhances the sense of presence (Cummings and Bailenson, 2016), and in binaural systems, significantly improves the quality of spatial audio rendering (Roginska, 2018).

The ability to shift viewpoint is, actually, essential in many augmented reality experiences (Schmalstieg and Hollerer, 2016). Holger Förterer's AAR installation *The Sound of Things* is based on the idea of exploring the scene with 6DoF, discussed in Chapter 5.4. Preceding narrative AR, the ability to change perspective has been common in interactive books throughout their long history (Rouse and Holloway-Attaway, 2020), and of course used in a number of video games where the player can find additional story content that deepens the 'lore' but does not necessarily affect the course of the main narrative.

Interactive storytelling enables different interpretations of the text and a unique experience for each viewer (McErlean, 2018). A shift of viewpoint, both contextual and physical, can dramatically change the interpretation of the narrative, especially if pre-planned by the author. On the other hand, since the experience is interactive, the user can decide not to change the perspective and rather experience the narrative on the surface level. With such a stationary viewpoint, the unobservable other side of the narrative element needs to be imagined and deduced based on expectations (see Smalley, 2007).

Crawford (2012) presents a concept of 'storyworld' where the storyteller is not feeding the participant with pre-planned events forming a 'picture-perfect view of the truth', but instead where the user can interact with ideas, observe them from various angles, and try different things: 'Storyworld presents truth in three dimensions, including the less elegant angles'. This is close to *environmental storytelling*, a concept prevalent in video game design, where stories are conveyed through information embedded in the environment. In narrative AAR experiences, the storyworld and environmental storytelling are useful methods, and can benefit from narrative techniques such as *attractor sounds* where objects emit sounds to draw user's attention to them (Zimmermann and Lorenz, 2008). Attractor sounds are further discussed in Chapter 9.

Although Crawford speaks metaphorically of the three-dimensional storyworld, one could draw an analogy to the built-in 3D nature of the AAR experience, especially with 6DoF. The player can move around, walk closer to sound sources, or further from them, allowing them to concentrate on the neighbours arguing behind the wall instead of listening to the main protagonist's monologue. Even if the virtual characters and objects in the storyworld cannot react to each action of the player, the change of perspectives alone may enable the player to experience and 'read' the story differently from other players. Temporal freedom, or ability to move and listen at their own pace, may also have a similar effect.

3.5 Acoustic ecology and soundscapes

Acoustic ecology is a field of study that explores how sounds shape our experience of the world and how they influence and reflect the ecological and cultural contexts of a place. It focuses on the relationship between living organisms and their acoustic environment (Schafer, 1977). Audio augmented reality is a way to re-examine that relationship and point out questions relating to human influence on the environment, or the environment's influence on individuals. Additionally, the concepts of acoustic ecology may also be useful when doing sound design for AAR experiences.

The key concept in acoustic ecology is *soundscape*, a term coined by R. Murray Schafer (1967; 1969). Soundscape describes the acoustic environment, the sonic equivalent of a landscape, comprising all the sounds within a particular area. Within a soundscape, one can identify different elements, such as *keynote sounds, signals, soundmarks,* and *hi-fi* and *lo-fi soundscapes* (Schafer, 1977). Keynote sounds are background sounds providing the underlying context of a place (e.g., nature sounds, the hum of a city). In acoustically isolated AAR displays, it is important to make sure these sounds are passed through or simulated in order to maintain the sense of presence as well as communicate the context of the place especially if visual sense is blocked. Signals, for their part, are foreground sounds, forcing conscious listening, such as bells, horns and sirens, whereas soundmarks are sonic landmarks, characteristic of a particular place and often holding cultural and historical significance, contributing to the identity of a community.

Hi-fi soundscapes have a favourable signal-to-noise ratio, where discrete sounds can be clearly heard (e.g., rural or natural environments), whereas lo-fi soundscapes are cluttered with noise, making it difficult to distinguish individual sounds (e.g., urban or industrial areas). Since soundscapes reflect the values, beliefs, and practices of a community, the transition from hi-fi to lo-fi soundscapes is one indicator of the cultural transformation from pre-industrial to industrial society: the loss of acoustic diversity and the dominance of noise suggests a shift in values from contemplation and connection with nature to speed and efficiency (Schafer, 1977).

Soundscapes also influence one's emotional and psychological state. This, consequently, affects the well-being and sense of place. With AAR, where the real-world soundscape is constantly omnipresent—unless technically attenuated—the influence of it must naturally be taken into account. For example, a geolocated audio walk created by an author with a deep, personal attachment to the soundscape of a specific environment may clash with a listener's entirely different relationship to the same auditory setting.

3.6 Environmental soundscape design

Because soundscapes hold such a big cultural, social, and environmental significance, urban planning should, as Barry Truax (2019) argues, draw from the principles of acoustic ecology and advocate for 'acoustic sustainability' to improve the quality of life. According to him, the sounds of built environments should remain on a human scale, using frequencies, loudness, spectral qualities, and temporal ranges similar to those produced by humans. He also acknowledges the role of sound artists and designers in creating virtual acoustic environments, either as simulations of future soundscapes to aid in design or as more abstract and intricate creations to foster engagement with the environment.

In urban planning and other infrastructure projects, AAR may be a useful tool to test and demonstrate the future soundscape in situ. For example, before planning a new residential area, several virtual acoustic versions of the predicted sounds could be constructed and played back during an AAR audio walk in the real environment. While simulations in a laboratory offer more control over distracting and unpredictable environmental factors, conducting the evaluations and hypotheses in the real environment potentially provides higher ecological validity (Hong et al., 2017).

In terms of technology, a mobile, binaural AAR system likely provides the most flexibility in this context, yet direct augmentation with loudspeakers placed within the environment would offer a straightforward and effective alternative for some cases (see Lacey, 2014). Although some experiments benefit acoustically transparent systems, acoustic isolation may also be useful when, for instance, assessing the effectiveness of noise-reduction measures.

3.7 Audio walks

Audio walks play an elemental role within the context of AAR. In an audio walk, the user traverses an area while hearing pre-recorded sounds from the same or some other place, forming narratives and compositional content synchronised with the environment (Papachristou, 2022). They can be seen as curated and enhanced versions of *soundwalks*, exercises where the intention is to listen with attention to the environment as it is (Westerkamp, 1974).

Audio walks, sometimes called 'audiotrails', are usually set in an outdoor environment, and they combine elements of public, local, and oral history, intersecting with various disciplines and art forms (Bradley, 2012). They are often realised by independent artists and attempt to create a strong bond between the personal accounts of the artist or the protagonists, the environment, and the participant. While the early audio walks—such as Cilia Eren's and Janet Cardiff's from the 1980s and 1990s (see Chapter 4.4)—were linear with a possibility to pause and rewind, geolocated audio walks have become popular since the 2010s. They enable automatic synchronisation of the content based on user's geographical position (see Chapter 5.6).

Audio walks invite participants to reflect on their relationship with the environment and explore the emotional impact of soundscapes. This idea draws from psychogeography, a framework introduced by the Situationist International in the 1950s, advocating for organic connections with the environment and criticising the commodification of urban spaces (Wollen, 1989). A key practice in psychogeography is *dérive*, or 'drift', which are unplanned journeys through public or semi-public places guided by sensory cues, encouraging participants to engage deeply with their environment and their soundscapes (Papachristou, 2022). AAR can be used to guide users on *dérive*-like explorations (e.g., Naphtali and Rodkin, 2019), providing narrative and other auditory cues that lead them through unexpected and intriguing parts of the city. These drifts could allow the users to reflect on their emotional and psychological relationship with their surroundings, fostering a deeper connection to the place and landscape.

3.8 Environmental sounds

As discussed earlier, hearing the aural environment is an important component of making the reality seem alive. To make AAR seamlessly integrate with the reality and create a sense of presence, this ability to hear the surroundings should be fostered.

3.8.1 Acoustic transparency

To be able to hear the environment, an acoustically transparent audio display can be used. In addition to maintaining the feel of presence, such a system should offer better externalisation of spatialised virtual sounds (McGill et al., 2020). Options for a transparent auditory display include open-back and open-ear headphones, headphones with electronic hear-through, cross-talk cancellation (CTC), and soundfield-based systems using loudspeakers. However, the drawback of acoustic transparency is that unpredictable, intrusive sounds can disrupt a carefully crafted sound design and narrative. These noises are challenging to integrate into a story and can compromise intelligibility. While high-quality spatial audio can help separate sounds from background noise, its effectiveness has limits.

The use of transparent or open-back headphones is often justified by their ability to maintain the user's awareness of their surroundings or connection with others. In practice, however, this may not always hold true: the close proximity of the transducers to the ears means that even small amounts of audio content can easily override or mask external sounds (Armbruster, 2024). This phenomenon also occurs in everyday life and even without headphones—someone speaking nearby can make it difficult to understand what another person is saying just a few metres away.

One aspect with unwanted sounds is that in AAR it may be difficult or impossible for the listener to know which sounds belong to the story and which do not. In cinema, we can usually distinguish sounds that are happening outside of the film's realm, and ignore them. In theatre, that distinction starts to get more blurry, especially with immersive theatre. In AAR, where everything around the participant can potentially be part of the experience, the challenge lies in preventing acoustic intrusions from being mistaken as part of the narrative—particularly when they coincidentally align with the story's themes or events.

One solution could be to make the narrative sounds more stylised, or in some other way to distinguish them from the audio already present in the environment (Dam et al., 2024). Another approach could be to craft the story so that it utilises the unpredictable nature of the real world, or at least tolerates it. Naturally, designing such experiences is not an easy task. Schütze and Irwin-Schütze (2018) propose one solution where the user can say 'Pardon?' whenever environmental noise masks the virtual audio content and reduces its intelligibility. In response, the system would repeat the content.

3.8.2 Control

The most straightforward solution to block out intrusive sounds is to use acoustically isolated, or closed-back headphones. This has been the chosen method for some AAR exhibitions such as *Maison Gainsbourg* (see Chapter 5.). This may lead to experiences becoming more focused and private, but at the same time, the externalisation of spatialised sounds may decrease, along with a diminished sense of presence and connection to the real environment. The lack of presence can be mitigated to some extent by introducing artificial background ambiences.

In an ideal situation, the designer has total control over which environmental sounds are incorporated, which are prioritised, and which should be blocked out (Naphtali and Rodkin, 2019). Technology can assist in this by enabling pseudo acoustic transparency, where microphones capture external sounds, and the system controls how much of them the listener hears (Albrecht, Lokki and Savioja, 2011). Even better would be an adaptive system that recognises and isolates the unwanted sounds and selectively erases them, based on the logic defined by the content designer. While semantic sound separation and cancellation systems are being developed (e.g., Veluri et al., 2023) and some functionalities already exist in noise-cancelling headphones and hearing aids, content creators may need to wait a while before such systems and APIs (application programming interfaces) are readily available. Selective pseudoacoustic systems would open a new world of possibilities in terms of aural manipulation of the environment.

It is good to remember, that background noise belongs to the world, and may even reinforce the reality-based experience. For example, Laurence Cliffe (2022) observed that in his AAR installation *Alien Encounters,* the noisy and busy environment led some visitors to kneel in front of audio-augmented radios to hear them better. Although that was not optimal for user experience, the noise prompted genuine interaction with the objects as if they had been real sound sources.

3.9 Conclusion

In 2016, when the American video game company Niantic released their *Pokémon GO* mobile game, the general public became aware of

Figure 3.4 A player is about to catch a Pokémon lurking in their neighbourhood
Source: Photograph by the author

the magic of AR when virtual and real co-existed: suddenly the animated, three-dimensional Pokémon were hanging around in everyone's own neighbourhood, if only perceivable through the tiny cellphone screen. What increased the sense of presence was the fact that one had to physically travel to places in order to visit PokéStops and Gyms.

However, in the game, the user's environment is not used as a gameplay element but merely as a backdrop, and players soon learned that the game was more efficient to play with the camera-based AR feature turned off—the AR overlay turning out to be, after all, a gimmick (Alha et al., 2023).

Similarly, within AAR games and narrative experiences, there are surprisingly many examples where the story or gameplay does not rely on the surrounding environment. Yet, one could argue that the true power and uniqueness of AAR lies in that very dependency—the ability to draw content from the real world around the user. In this way, the application goes beyond merely leveraging sensorimotor interaction, such as requiring the user to move or gesture within an arbitrary environment. Instead, it overlays a narrative that comments on and reflects the specific place and the narrative layers it has accumulated over time. That would be the superpower of AAR.

References

Albrecht, R., Lokki, T. and Savioja, L. (2011) 'A Mobile Augmented Reality Audio System with Binaural Microphones', in *Proceedings of Interacting with Sound Workshop: Exploring Context-Aware, Local and Social Audio Applications.* New York, NY, USA: Association for Computing Machinery (IwS '11), pp. 7–11. Available at: https://doi.org/10.1145/2019335.2019337.

Alha, K. et al. (2023) 'Augmented Play: An Analysis of Augmented Reality Features in Location-Based Games', *Convergence*, 29 (2), pp. 342–361. Available at: https://doi.org/10.1177/13548565231156495.

Armbruster, S. (2024) 'Conversation with Matias Harju, March 7'.

Auslander, P. (2022) *Liveness: Performance in a Mediatized Culture.* 3rd edn. London: Routledge. Available at: https://doi.org/10.4324/9781003031314.

Azuma, R. et al. (2001) 'Recent Advances in Augmented Reality', *IEEE Computer Graphics and Applications*, 21 (6), pp. 34–47. Available at: https://doi.org/10.1109/38.963459.

Blauert, J. (1997) *Spatial Hearing : The Psychophysics of Human Sound Localization.* Rev. edn. MIT Press.

Bradley, S. (2012) 'History to Go: Oral History, Audiowalks and Mobile Media' (pre-published version). Available at: https://www.academia.edu/1148470/History_to_go_oral_history_audiowalks_and_mobile_media_pre_published_version_ (Accessed: 1 August 2024).

Breunig, D. (2024) 'The Rise of Audio Augmented Reality', *Medium.* Available at: https://medium.com/design-bootcamp/the-rise-of-audio-augmented-reality-c56a6348ff59 (Accessed: 20 November 2024).

Cliffe, L. (2022) *Audio Augmented Objects and the Audio Augmented Reality Experience.* PhD thesis. University of Nottingham. Available at: https://eprints.nottingham.ac.uk/69795/.

Cliffe, L. (2024) 'Into the Here and Now: Explorations within a New Acoustic Virtual Reality', *Leonardo* [Preprint]. Available at: https://nottingham-repository.worktribe.com/output/4248043/into-the-here-and-now-explorations-within-a-new-acoustic-virtual-reality (Accessed: 2 December 2024).

Crawford, C. (2012) *Chris Crawford on Interactive Storytelling*, 2nd edn. Available at: https://learning.oreilly.com/library/view/chris-crawford-on/9780133119671/ (Accessed: 17 January 2021).

Csikszentmihalyi, M. (1990) 'Literacy and Intrinsic Motivation', *Daedalus*, 119 (2), pp. 115–140.

Cummings, J.J. and Bailenson, J.N. (2016) 'How Immersive Is Enough? A Meta-Analysis of the Effect of Immersive Technology on User Presence', *Media Psychology*, 19 (2), pp. 272–309. Available at: https://doi.org/10.1080/15213269.2015.1015740.

Cutler, K.-M. (2009) 'The Future of the Music Album? Check Out RjDj's Little Boots App', *VentureBeat*, 16 December. Available at: https://venturebeat.com/business/rjdj/ (Accessed: 24 October 2024).

Dam, A. et al. (2024) 'Audio Augmented Reality Using Sonification to Enhance Visual Art Experiences: Lessons Learned', *International Journal of Human-Computer Studies*, 191, p. 103329. Available at: https://doi.org/10.1016/j.ijhcs.2024.103329.

Fendt, M.W. et al. (2012) 'Achieving the Illusion of Agency', in D. Oyarzun et al. (eds) *Interactive Storytelling*. Berlin, Heidelberg: Springer, pp. 114–125. Available at: https://doi.org/10.1007/978-3-642-34851-8_11.

Glassner, A. (2017) *Interactive Storytelling: Techniques for 21st Century Fiction*. CRC Press.

Green, M.C. (2021) 'Transportation into Narrative Worlds', in L.B. Frank and P. Falzone (eds) *Entertainment-Education Behind the Scenes*. Palgrave Macmillan, Cham, pp. 87–101. Available at: https://doi.org/10.1007/978-3-030-63614-2_6.

Hong, J.Y. et al. (2017) 'Spatial Audio for Soundscape Design: Recording and Reproduction', *Applied Sciences*, 7 (6), p. 627. Available at: https://doi.org/10.3390/app7060627.

Järvinen, A. (2017) 'Design for Presence in VR, Part 2: Towards An Applied Model ', *Medium*. Available at: https://virtualrealitypop.com/designing-for-presence-in-vr-part-2-towards-an-applied-model-2784bf16a01 (Accessed: 29 November 2024).

Kuhn, G. (2019) *Experiencing the Impossible: The Science of Magic*. The MIT Press. Available at: https://doi.org/10.7551/mitpress/11227.001.0001.

Lacey, J. (2014) 'Site-Specific Soundscape Design for the Creation of Sonic Architectures and the Emergent Voices of Buildings', *Buildings*, 4 (1), pp. 1–24. Available at: https://doi.org/10.3390/buildings4010001.

Larsson, P. et al. (2010) 'Auditory-Induced Presence in Mixed Reality Environments and Related Technology', in E. Dubois, P. Gray, and L. Nigay (eds) *The Engineering of Mixed Reality Systems*. London: Springer (Human-Computer Interaction Series), pp. 143–163. Available at: https://doi.org/10.1007/978-1-84882-733-2_8.

Lombard, M. and Ditton, T. (1997) 'At the Heart of It All: The Concept of Presence', *Journal of Computer-Mediated Communication*, 3(JCMC321). Available at: https://doi.org/10.1111/j.1083-6101.1997.tb00072.x.

McErlean, K. (2018) *Interactive Narratives and Transmedia Storytelling: Creating Immersive Stories Across New Media Platforms*. 1st edn. Milton: Routledge. Available at: https://doi.org/10.4324/9781315637570.

McGill, M. et al. (2020) 'Acoustic Transparency and the Changing Soundscape of Auditory Mixed Reality', in *Proceedings of the 2020 CHI Conference on Human Factors in Computing Systems*. Honolulu, Hawaii, USA: ACM, pp. 1–16. Available at: https://doi.org/10.1145/3313831.3376702.

Merleau-Ponty, M. (2013) *Phenomenology of Perception*. London: Routledge. Available at: https://doi.org/10.4324/9780203720714.

Milgram, P. and Kishino, F. (1994) 'A Taxonomy of Mixed Reality Visual Displays', *IEICE Transactions on Information and Systems*, E77-D(12), pp. 1321–1329.

Naphtali, D. and Rodkin, R. (2019) 'Audio Augmented Reality for Interactive Soundwalks, Sound Art and Music Delivery', in *Foundations in Sound Design for Interactive Media*. Routledge.

Papachristou, D. (2022) 'Locative Media Walks: Geo-Locative Media as a Means of Subverting Hegemonic Historiography', *Revista SOBRE*, 8, pp. 31–40. Available at: https://doi.org/10.30827/sobre.v8i.23875.

Roginska, A. (2018) 'Binaural Audio Through Headphones', in A. Roginska and P. Geluso (eds) *Immersive Sound: The Art and Science of Binaural and Multi-Channel Audio*. Routledge.

Rouse, R. and Holloway-Attaway, L. (2020) 'A prehistory of the interactive reader and design principles for storytelling in postdigital culture', *Book 2.0*, 10 (1), pp. 7–42. Available at: https://doi.org/10.1386/btwo_00018_1.

Schafer, R.M. (1967) 'Ear Cleaning: Notes for an Experimental Music Course'. BMI Canada. Available at: https://monoskop.org/images/2/2d/Schafer_R_Murray_Ear_Cleaning_Notes_for_an_Experimental_Music_Course.pdf (Accessed: 9 October 2024).

Schafer, R.M. (1969) *The New Soundscape: A Handbook for the Modern Music Teacher*. BMI Canada.

Schafer, R.M. (1977) *The Tuning of the World*. New York: A. A. Knopf. Available at: http://archive.org/details/tuningofworld0000scha (Accessed: 9 October 2024).

Schmalstieg, D. and Hollerer, T. (2016) *Augmented Reality: Principles and Practice*. Addison-Wesley Professional.

Schraffenberger, H. (2018) *Arguably Augmented Reality: Relationships Between the Virtual and the Real*. PhD thesis. Universiteit Leiden. Available at: https://www.creativecode.org/wp-content/uploads/Thesis/thesis-print.pdf (Accessed: 14 November 2024).

Schütze, S. and Irwin-Schütze, A. (2018) *New Realities in Audio: A Practical Guide for VR, AR, MR and 360 Video*. CRC Press. Available at: https://www.routledge.com/New-Realities-in-Audio-A-Practical-Guide-for-VR-AR-MR-and-360-Video/Schutze-Irwin-Schutze/p/book/9781138740815 (Accessed: 30 January 2024).

Siirtola, P. et al. (2023) 'Predicting Emotion with Biosignals: A Comparison of Classification and Regression Models for Estimating Valence and Arousal Level Using Wearable Sensors', *Sensors*, 23 (3), p. 1598. Available at: https://doi.org/10.3390/s23031598.

Slater, M. (2003) 'A Note on Presence Terminology', *Presence Connect*, 3 (3), pp. 1–5.

Smalley, D. (2007) 'Space-form and the Acousmatic Image', *Organised Sound*, 12 (1), pp. 35–58. Available at: https://doi.org/10.1017/S1355771807001665.

Stevens, M. (2009) *Music and Image in Concert: Using Images in the Instrumental Music Concert*. Sydney: Music and Media.

Truax, B. (2019) 'Acoustic Sustainability in Urban Design: Lessons from the World Soundscape Project', *Cities & Health*, 5(1–2), pp. 122–126. Available at: https://doi.org/10.1080/23748834.2019.1585133.

Turner, C. (2022) 'Augmented Reality, Augmented Epistemology, and the Real-World Web', *Philosophy & Technology*, 35 (1), p. 19. Available at: https://doi.org/10.1007/s13347-022-00496-5.

Veluri, B. et al. (2023) 'Semantic Hearing: Programming Acoustic Scenes with Binaural Hearables', in *Proceedings of the 36th Annual ACM Symposium on*

User Interface Software and Technology. New York, NY, USA: Association for Computing Machinery (UIST '23), pp. 1–15. Available at: https://doi.org/10.1145/3586183.3606779.

Verhagen, T. et al. (2014) 'Present It Like It Is Here: Creating Local Presence to Improve Online Product Experiences', *Computers in Human Behavior*, 39, pp. 270–280. Available at: https://doi.org/10.1016/j.chb.2014.07.036.

Wakkary, R. et al. (2004) 'The Use of Audio for a Dynamic Museum Experience through Augmented Audio Reality and Adaptive Information Retrieval', in *Museums and the Web 2004: Selected Papers from an International Conference.* Vancouver, Canada, pp. 55–60. Available at: http://summit.sfu.ca/item/15157.

Westerkamp, H. (1974) 'Soundwalking', *Sound Heritage*, pp. 18–27.

Wirth, W. et al. (2007) 'A Process Model of the Formation of Spatial Presence Experiences', *Media Psychology* [Preprint]. Available at: https://doi.org/10.1080/15213260701283079.

Wollen, P. (1989) 'The Situationist International', *New Left Review*, 1 (174), pp. 67–95.

Yang, J., Barde, A. and Billinghurst, M. (2022) 'Audio Augmented Reality: A Systematic Review of Technologies, Applications, and Future Research Directions', *Journal of the Audio Engineering Society*, 70 (10), pp. 788–809.

Zimmermann, A. and Lorenz, A. (2008) 'LISTEN: A User-Adaptive Audio-Augmented Museum Guide', *User Modeling and User-Adapted Interaction*, 18 (5), pp. 389–416. Available at: https://doi.org/10.1007/s11257-008-9049-x.

4   From echoes to audio walks

Sonic augmentations, whether deliberate or incidental, have likely existed for as long as there have been ears capable of perceiving them. Ghostly sounds heard inside a prehistoric cave have offered a layer to an alternative reality coexisting with the tangible one. It is this 'magical' blending of virtual sounds with the real environment that has long captivated the imagination and creativity of both creators and listeners.

The next two chapters will explore the historical paths leading to the current-day AAR through example projects and significant milestones. This chapter begins with a brief look into prehistory and progresses through the 20th century, concluding with radio-based *locative* audio walks from the turn of the millennium—precursors, or creative alternatives, to modern geolocated applications. (For the term locative, see Chapter 9.1.) While these examples were likely not identified as AAR at the time, they share its core principle: mixing artificial audio with reality to generate new meanings. Due to space constraints, the focus is placed on practical applications and experiments, with less emphasis on equally significant scientific research and product development.

4.1 Talking spirits and dummies

The concepts of phantom sounds appearing in the real environment and inanimate objects magically producing sounds are by no means new. It is believed that acoustic illusions have been a significant inspiration for many prehistoric cave paintings: the image of a spirit drawn at the very spot where the artist's voice has been echoed from the cave wall, as if the supernatural spirit were talking to the artist from inside the solid rock (Waller, 2002).

In the Chaldean temples (ca. 900–500 BCE) in Mesopotamia, the priests questioned their idols, the theraphim, whose statues magically 'answered' in the presence of the kneeling people; the trick was to connect a windpipe of a long-necked bird to the head of the theraphim dummy and have an

DOI: 10.4324/9781003627289-4

accomplice speak into the pipe, making the head appear to talk (Pettorino, 2015). Similar illusions using concealed voice pipes and cones have since been used throughout centuries by magicians and other tricksters.

In the aforementioned examples, the physical sound waves were actually bouncing or emanating from their 'virtual' locations. Ventriloquism, on the other hand, is an example of an ancient practice where the sound only *appears* as coming from another location; when a ventriloquist speaks without moving their lips and simultaneously manipulates the mouth of a puppet, the spectator perceives the voice coming from the dummy. This phenomenon is based on *visual capture*, or visual domination over auditory perception (e.g., Pick, Warren and Hay, 1969). This is called the *ventriloquism effect* (Howard and Templeton, 1966). Visual capture is not, however, absolute; while visual information significantly influences sound localisation, it does not entirely override auditory perception (Majdak, Goupell and Laback, 2010).

In headphone-based AAR, the ventriloquism effect plays an important role in helping virtual sounds feel attached to real-world objects even with imperfect audio spatialisation or externalisation. Michel Chion (1994) refers to this effect as 'magnetisation' and provides a common example: the experience of watching a video with headphones on while still perceiving the sound as coming from the screen.

4.1.1 Ear trumpets and sound locators

The spirits on cave walls, the talking statues, and the ventriloquist's dummy can be experienced with naked ears without any additional aids. However, people have also boosted—or augmented—their senses in order to hear the inaudible.

Ear trumpets have likely been used as hearing aids since ancient times. Without essential changes in the design, they were widely used throughout centuries and even millennia up until the 1970s when electronic hearing aids replaced them (Briggs, 2023). Ear trumpets were usually held in one ear only, providing monaural auditory augmentation that was sufficient for surviving social situations. Nothing prevented, however, using two trumpets for binaural augmentation. When using large, artificial ears, remote sounds are amplified while their rough location can be perceived using binaural cues. Such devices were manufactured in all sizes from personal 'elephant ears' to huge 'sound tubas' operated by multiple personnel. They have been used since, at least, the 19th century for maritime audio-location in bad visibility, and they became popular during World Wars I and II for anti-aircraft detection and location before the radar was invented in the 1930s (Self, 2024).

Figure 4.1 A Dutch personal binaural acoustic locator presumably from the 1930s
Source: Public domain (rarehistoricalphotos.com)

4.1.2 Binaural audio

Speaking tubes and voice pipes have been used for centuries for communication in buildings and maritime vessels (Briggs, 2023), allowing telepresence over tens or maybe hundreds of metres. Telephone, invented in the 1870s, and radio with first transmissions in 1890 revolutionised long-distance communication. One of the goals from early on was to recreate the soundscape and appearance of the transmitting party as accurately as possible at the receiver's end. Using pairs of telephone transmitters and receivers, Bell himself conducted such 'binaural' experiments as early as in 1880 (Davis, 2003), followed by Clement Ader the following year (Boren, 2018).

Bell's influence on binaural audio continued at the Bell Labs where multichannel audio was researched under the direction of Harvey Fletcher, leading to one of the earliest stereophonic wax disk recordings in 1932 with the Philadelphia Orchestra (Schoenherr, 1999) as well as a three-channel transmission 200 km away to Chicago, foreshadowing wave field synthesis and paving the way for surround sound systems (Boren, 2018).

The Bell Labs also developed one of the earliest binaural dummy heads named 'Oscar', demonstrated at the Chicago World's Fair in 1933 to the amazement and approval of the audience (Boren, 2018). Even though radio stations and record companies experimented with binaural techniques especially in the 1970s (Bülow, 2013; Berry, 2017), binaural listening never took off with a wider audience, possibly because the consumers, predating the upcoming Sony Walkman boom of the 1980s, were not yet accustomed to using headphones. However, in virtual audio research and applications such as AAR, the use and development of binaural technologies have remained central.

4.1.3 Edison's endeavours

One of the most important early inventions in terms of AAR, and virtual sound in general, is sound recording. The idea of capturing sound 'in a tube' and playing it on request had been around at least since the 16th century (Pettorino, 2015). It took, however, three centuries before the idea became technically possible. While Martinville's phonautograph, invented in 1857, was the first ever sound recording apparatus, it could not play back audio. The phonograph, patented by Thomas Edison in 1877, however, did just that, and for the first time in human history it was possible to capture real sound and play it back (Dawson, 2020).

Edison, however, struggled to find customers for his invention, and one desperate attempt to find new markets for the phonograph was the talking doll (Gronow and Saunio, 1990). These were manufactured only for a short period of time in 1890, equipped with a tiny wax cylinder etched with a 20-second recording of a nursery rhyme such as 'Twinkle, Twinkle, Little Star' and 'Mary Had a Little Lamb' (Dawson, 2020). This business endeavour, too, quickly turned out to be a commercial flop due to the fact that these expensive toys were easy to break apart and the sound quality was not just poor but made the dolls sound creepy. Nevertheless, Edison's talking dolls remain as some of the earliest examples of using direct audio augmentation of a physical, real-world object.

In 1912, however, the time was ripe for recorded audio—mostly music—and there were a number of competing manufacturers of play-back-only disk phonographs, or gramophones as they started to be called. Edison himself attempted to take a slice of this booming market with the release of his 'New Edison' phonograph. The machine used the proprietary 'Diamond Disc' format and was unpractical and heavy (Apple, 2023; Gronow and Saunio, 1990). Nevertheless, it had a superb sound quality, at least as advertised: it could 'Re-Create' the tone quality of the original sounds and voices 'absolutely' without hearable difference (Thomas A. Edison Inc., 1920).

Between 1915 and 1925, Edison arranged a wide *Tone Test* marketing campaign where famous opera singers such as Anna Case and other classical musicians performed on stage beside the phonograph, and the audience or an expert jury tried to tell when it was the musician and when the phonograph that produced the music. One trick was to let the singer start a song and then switch off the lights. When the lights came back on, the singer had disappeared, but the music was still continuing uninterrupted, now emanating from Edison's phonograph. It is unclear how many unbiased reports have survived from the actual audience members, but reportedly, most audience members could not tell the difference between the real musician and the phonograph (Apple, 2023).

Whereas the talking doll suffered from technical limitations and failed to create an auditory illusion, this time, Edison really attempted to 'virtualise' a sound using his phonograph and spatially (nearly) align it with the contextually congruent real-world counterpart—the musician or their instrument—to create a plausible illusion and audio augmentation.

4.2 Speaker-based direct augmentations

Edison's talking doll and the performances during the Tone Test campaign were probably the first examples of direct augmentation using *recorded*

Figure 4.2 A picture from a magazine advertisement showing a jury behind a folding screen, attempting to distinguish the violin from the 'New Edison' phonograph recording. *Saturday Evening Post*, August 14, 1920
Source: Photograph provided by Curtis Licensing

sound, a significant step towards AAR. The development of sound recording, playback systems and electronic loudspeakers since the 1920s have paved the road for direct audio augmentation spreading to museums, attractions and other venues all over the world.

One relatively early example of extensive use of this approach is the original Disneyland in California. When opened in 1955, it was full of speaker-based direct augmentations. For instance, the *Jungle Cruise* presented static animals accompanied by animal and jungle sounds played through hidden speakers. In 1963, the park introduced the *Enchanted Tiki Room* filled with electromechanical puppets trademarked as 'Audio-Animatronics' that featured synchronised audio through concealed speakers (Kroon, 2010). Many more attractions followed, both in Disneyland and other amusement parks around the world with similar concepts still in use in current day exhibitions.

Figure 4.3 Talking parrot Pierre, an 'Audio-Animatronic', in the *Enchanted Tiki Room* at Disneyland
Source: Photograph by Anna Fox/CC BY

4.2.1 Public spaces

Hidden speakers in public spaces is also a popular form of sound art. One famous example of such is *Times Square* by the American sound artist Max Neuhaus. The installation, created in 1977, is located on a traffic island on Broadway in Manhattan and is still running at the time of writing this. The piece is not signposted but found by hearing 'a rich, ringing drone, like a deep, industrial hum' emanating under the grates of a subway ventilation system (Kotz, 2009, p. 93). Neuhaus attempted to design the sound to be 'plausible', something that could be mistaken for a sound emanating from the subway system (Dia Art Foundation, 2023). We can, therefore, place the piece within the framework of AAR even with the lack of interactivity. This intentional ambiguity of blurring of art and environment, or virtual and real, is something Neuhaus sought for, prompting listeners to question their perception of the environment (Dia Art Foundation, 2023).

Volkswagen's advertisement campaign *The World's Deepest Bin* in 2009 is another clever example of hidden speakers in a public space. A litter bin at a park in Stockholm, Sweden, was equipped with a motion sensor and battery-operated speakers. When a passer-by threw a piece of litter inside it, a cartoonish sound of an ascending whistle was heard, as if the item was falling hundreds of metres below the surface of the earth until a reverberant hit sounded from the bottom of the 'world's deepest bin'. This fun idea made people not just dispose of their litter properly but also pick up other people's trash just to hear the sound (Yu, 2022).

In terms of AAR, the installation was largely based on *extension* of the space (see Chapter 9.8), suggesting an extended virtual space far beyond the bin's physical dimensions. This auditory illusion was combined with the exaggerated and culturally nostalgic cartoon sound trope that further took the extension effect beyond reality with an added element of absurdity.

Audio augmentations have been taken into nature, too. In 2019, researchers from University of Cumbria and Manchester Metropolitan University, UK (Lawton, Cunningham and Convery, 2020) installed four concealed speakers in a forest in order to explore the potential of AAR to enhance audience engagement with and understanding of nature. In this *Nature Soundscapes* project, they played back sounds of current and extinct species to raise awareness about species decline and human impact on bio-diversity. Also, sounds of human activity and environmental degradation were superimposed on the forest soundscape, emphasising the impact of human-made sounds on nature. The loudspeaker approach was chosen because it intrinsically allowed participants to hear the real forest as it is, potentially enabling them to grasp the juxtapositions more intuitively.

4.2.2 *Museums:* Barque Sigyn

Hidden loudspeakers have been popular in museums because they provide a simple and reliable way to transport the exhibition or individual exhibits into another era. *Barque Sigyn* is an example of narrative augmentation in a museum setting with simple interactivity. Sigyn is a preserved sailing vessel from 1887 and opened to the public in 2021 at the Turku maritime museum in Finland. At one point, when the guided tour reaches the captain's cabin, the visitor can hear, somewhere around the corner, something metallic drop and a cat meow. The illusion is plausible, as if there really was a cat somewhere, even though that would be somewhat unlikely in a museum setting. As the tour moves to the back room, the source of the sounds is cleverly revealed using a simple prop (without giving away too much). Even though the realisation is as simple as it gets—a motion detector, a single sound clip, and a hidden speaker (Niittymäki, 2024)—it successfully builds a tiny but concise narrative, synchronised with the visitor's location, that not only sparks amusement but also briefly immerses them in the once-lived life aboard the ship.

Figure 4.4 Captain's cabin onboard *Barque Sigyn* with a virtual cat sound coming from the next room
Source: Photograph by the author

4.2.3 Cars

Modern cars, especially electric ones, are an emerging field for speaker-based audio augmentations. Electric vehicles are inherently nearly silent, posing a safety risk for pedestrians, especially children and visually impaired individuals. This has led regions like the EU, USA, Japan, and China to mandate *acoustic vehicle alerting systems* (AVAS), which use exterior speakers to produce artificial sounds at low speeds (Thor, 2024).

Further, inside the cabin, virtual engine sounds may enhance the driving experience. Some cars, like the Ford Mustang Mach-E, let drivers choose from a selection of sounds, which attempt at providing the sense of horse power and torque without emitting noise pollution. The design of these artificial exterior and interior sounds often blend traits from combustion engines with a futuristic, space-ship like aesthetic.

Loudspeakers are also used for active noise cancellation (ANC) to reduce road and engine noises, using inverted sound waves to counter low-frequency sounds more effectively than traditional acoustic treatments (e.g., Ford, 2015; Hyundai, 2020; Bose, 2024; Harman, 2024). Additionally, spatialised sonification through surround speakers can convey situational awareness information, such as obstacle alerts, by localising sounds in the direction of the detected object.

4.3 Headphones

The invention of electric headphones and earbuds, both in the 1890s (Stamp, 2023), was crucial for the development of AAR. They enabled a personal listening experience, isolation of environmental sounds, and binaural playback of transmitted and virtual soundscapes.

The military obviously became a major user of headphones, and they also utilised them binaurally: for example, during World War I, some sound locators were equipped with headphones (Self, 2024), and submarines were tracked using the binaural auditory image from two hydrophones (Boren, 2018). The Luftwaffe used an innovative blind navigation system in World War II based on two-channel headphone listening: one strategically located radio station sent Morse dots to the pilot's left headphone while another station sent dashes to the right ear exactly in between the dots. A deviation from the assigned course would off-sync the beeps, thus prompting the pilot to make a correction (Kittler, Winthrop-Young and Wutz, 1999).

Further, research on active noise control since the 1930s and the first commercial ANC headphones by Bose in 1989 mark significant milestones for AAR as noise reduction later became the first AAR feature widely available for and widely used by consumers. This has enabled dynamic manipulation of one's personal soundscape, a long step from merely controlling sound levels.

4.4 Audio guides and audio walks

Headphones and earbuds enabled the concepts of audio walks and audio guides. While the early implementations lacked the location-based inter-activity of more recent applications, their content was strongly related to the particular location—or intentionally divergent from it.

4.4.1 Audio guides

Audio guides are mediated versions of guided tours. They lack the presence of a human guide, but enable the experiencing of the exhibition with a more flexible or even personal schedule, in multiple languages, without obtruding fellow patrons—and are cheaper to run than hiring human guides. In the early applications, audio content was linear with sometimes a possibility to pause and rewind. In the 1980s, with infrared and radio technology, individual exhibits could be broadcasting looped audio when the visitor came close enough, often arriving in the middle of the loop and thus missing the start of the description. Later, museums implemented systems where the user tapped a code to choose correct tracks from a random access audio storage device (Eckel, 2001), and slowly automatic locative systems started to emerge in the 2000s. Yet, even now, many modern audio guides and walks are not capable of head tracking that would enable spatial alignment of the auditory content with the environment. Still, in a well-designed audio guide, even with linear content, the audio merges with the environment and invites 'the participant to make connections between the added sound and the existing environment' (Schraffenberger and Heide, 2014, p. 68). That has been a solid foundation to build upon for more recent locative and 6DoF experiences.

One of the earliest known implementations of an audio guide was introduced in 1952 at the Stedelijk Museum in Amsterdam. The linear voice description was recorded onto a magnetic tape and broadcast inside the museum using an induction loop hidden behind the skirting boards; visitors participating in the tour wore an earpiece and picked up the radio signal by a small receiver (Nederlands Instituut voor Beeld and Geluid, 2020). This pioneering 'wireless tour' aroused interest in several museums (Stedelijk Museum Amsterdam, 2021), and a similar system was later in use at the National Gallery of Washington (Kennedy, 2006).

Figure 4.5 Visitors on the *Alcatraz Cellhouse Audio Tour*
Source: Photograph by JiaYing Grygiel

A current-day audio guide example is the widely recognised *Alcatraz Cellhouse Audio Tour* in the legendary maximum security federal prison in the San Fransisco Bay that was turned into a museum in 1980s. The original audio tour was made in 1987 using Sony Walkmans, and during the renovation in 2007 a new version was made with redesigned sound effects and MP3 audio players (Golden Gate National Parks Conservancy, 2007).

The 45-minute audio presentation features the actual voices of correctional officers and inmates who lived and worked on the island during its prison era. Atmosphere is created through the clanging of cell doors, footsteps, and other prison sounds. Instead of using numbered locations where one has to press buttons to trigger audio for individual sections, the audio tour plays back a linear presentation that assumes that visitors will follow the prescribed path at a relatively consistent pace, similar to the Stedelijk Museum guide. If a visitor lingers in an area or moves faster than expected, they can pause and start the audio manually. That offers a low-maintenance solution without the need to install any tracking infrastructure on the historical structures.

Like the Stedelijk system, the audio at Alcatraz is head-locked, albeit still able to create immersive moments when, for instance, the visitor is standing in a corridor and hears the banging of the bars and shouts from the inmates around. At some points, the visitor is invited to synch their steps with one of the inmates, a similar technique to foster engagement with the protagonist as used by Janet Cardiff in some of her audio walks.

4.4.2 Audio walks

Janet Cardiff and her partner–collaborator George Bures Miller have been significant figures in the development of audio walks as a form of site-specific sound art. In Cardiff's *Lousiana Walk #14* from 1996, the visitor of an art gallery hears precise audio instructions through headphones, guiding them out of the indoor space 'to start an outdoor adventure' (Butler, 2007). The use of binaural dummy head recordings in Cardiff's audio walks, started from the *Münster Walk* in 1997 (Cardiff and Miller, 2024), create immersion and co-presence with the protagonist; while walking on the same streets, the participant is also encouraged to follow the pace of the narrator's footsteps (Butler, 2007). The walks often mix multi-layered fictional narratives with the reality and physical movement through it (Bradley, 2012), thus experimenting with the core elements of AAR.

The first Cardiff–Miller audio walks used portable CD players and iPods with linear content and possibility to pause and rewind if necessary, to be replaced by smart phones by the end of the first decade of the 2000s (Bradley, 2012).

Audio walks by the recently passed-away Dutch sound artist Cilia Eren were quite the opposite to Cardiff's in the sense that, firstly, Eren was never audibly present but gave the participant the protagonist's role, and secondly, the idea was to dislocate the participant: for example, in her first audio walk *China Daily* from 1987, the listener was invited to imagine being in a busy Chinese city while wandering the streets of Amsterdam (Steijn, 2023). While such acoustic translocation is usually difficult to realise, the narrative transportation and the common modality of walking in both environments—auditory and physical ones—may be of help.

4.4.3 Radio-based locative experiences

The introduction of satellite positioning and mobile internet into people's wearable devices enabled the development of geolocated audio walks and experiences during the first decade of the 2000s, to be discussed in the next chapter. Before that, however, some sound artists used radio transmitters and receivers to create locative experiences with environment-embedded virtual sound sources.

In 2000, Edwin van der Heide (2023) arranged his first *Radioscape* installation that explores radio as a spatial and interactive medium. Fifteen custom transmitters scattered throughout an urban environment broadcast layers of a 'meta-composition', which participants hear using a specially designed receiver. The receivers have a couple of interesting features. First, they are able to pick up multiple signals simultaneously. Secondly, the distance between the transmitter and receiver directly affects the loudness even on a short distance, something that is uncommon for many people used to listening to commercial and public radio stations transmitting with high power.

Thirdly, and perhaps most interestingly, the receivers are 'stereo panoramic receivers' as the artist calls them. By utilising two antennas and stereo headphones, they act like a stereo microphone: the participants hear the direction and distance of the transmitters; stations to the left of the antenna are heard more on the left ear, and vice versa.

This rare ability to hear radio signals spatially makes the transmitters, in a way, virtual sound sources embedded in the environment. The way radio waves interact with urban structures adds further complexity, with reflections and interference altering the soundscape, similar to how sound waves behave when propagating through an environment. All

Figure 4.6 Participants of *Radioscape* using the custom-made radio receivers
Source: Photograph by Edwin van der Heide, courtesy of Studio Edwin van der Heide

these combine with the participants' ability to explore—not just the urban environment but—the radiophonic composition and create different mixes based on their location, movement and orientation, the installation connecting tightly to the context of audio augmented reality with the flavour of 'super hearing' (as discussed in Chapter 2.6).

Another radio-based audio experience is *Linked*, installed in 2003 in East London, UK, by Graeme Miller (2024). The installation utilises multiple radio transmitters that create 'geolocated' reception zones. The original work was commissioned by the Museum of London as a response to the creation of the M11 Link Road, which displaced over 400 buildings and affected a vibrant community. Along a 5 km route along the road, several transmitters installed in lampposts deliver an audio trail of memories, stories, and voices from the area's past residents, including artists, workers, and activists involved in passionate protests against the development.

In 2003, mobile audio walk apps using satellite positioning were not yet available to the public. The use of radio transmitters was an elegant predecessor, however requiring physical installation, power-supplies, and maintenance. Like with Heide's work, physical radio transmitters concretely root the content in the environment. *Linked* uses audio loops, which have the disadvantage that listeners picking up a signal mid-loop may need to wait for the next story to begin (Bederson, 1995). In *Linked*, however, this together with the intermittent radio signal reception, creates a feeling of nostalgia suitable for the piece, further allowing multiple participants to be in perfect synch with the content, something that is challenging to realise with web-based systems (Bradley, 2012).

References

Apple, R.W. (2023) 'Turning Wax into Diamonds', *Edison and Ford Winter Estates*, 1 December. Available at: https://www.edisonfordwinterestates.org/turning-wax-into-diamonds/ (Accessed: 2 July 2024).

Bederson, B.B. (1995) 'Audio Augmented Reality: A Prototype Automated Tour Guide', in *Conference Companion on Human Factors in Computing Systems – CHI '95*. Denver, Colorado, United States: ACM Press, pp. 210–211. Available at: https://doi.org/10.1145/223355.223526.

Berry, G. (2017) 'History of Binaural'. Available at: https://geberry.wixsite.com/binaural/history (Accessed: 10 September 2024).

Boren, B. (2018) 'History of 3D Sound', in A. Roginska and P. Geluso (eds) *Immersive Sound: The Art and Science of Binaural and Multi-Channel Audio*. Routledge, pp. 40–62.

Bose (2024) 'Active Sound Management | Bose Automotive'. Available at: https://automotive.bose.com/road-ahead/active-sound-management (Accessed: 25 September 2024).

Bradley, S. (2012) 'History To Go: Oral History, Audiowalks and Mobile Media' (pre-published version). Available at: https://www.academia.edu/1148470/History_to_go_oral_history_audiowalks_and_mobile_media_pre_published_version_ (Accessed: 1 August 2024).

Briggs, M.J. (2023) 'A Brief History of the Acoustic Ear Trumpet and Some Collection Favourites', *ENT & Audiology News*, December. Available at: https://www.entandaudiologynews.com/features/audiology-features/post/a-brief-history-of-the-acoustic-ear-trumpet-and-some-collection-favourites (Accessed: 15 September 2024).

Bülow, R. (2013) 'Vor 40 Jahren: Ein Kunstkopf für binaurale Stereophonie', *heise online*. Available at: https://www.heise.de/news/Vor-40-Jahren-Ein-Kunstkopf-fuer-binaurale-Stereophonie-1946286.html (Accessed: 10 September 2024).

Butler, T. (2007) 'Memoryscape: How Audio Walks Can Deepen Our Sense of Place by Integrating Art, Oral History and Cultural Geography', *Geography Compass*, 1 (3), pp. 360–372. Available at: https://doi.org/10.1111/j.1749-8198.2007.00017.x.

Cardiff, J. and Miller, G.B. (2024) 'Münster Walk, Janet Cardiff & George Bures Miller'. Available at: https://cardiffmiller.com/walks/munster-walk/ (Accessed: 31 October 2024).

Chion, M. (1994) *Audio-Vision: Sound on Screen*. Columbia University Press.

Davis, M. (2003) 'History of Spatial Coding', *Journal of the Audio Engineering Society*, 51 (6), pp. 554–569.

Dawson, V. (2020) 'The Epic Failure of Thomas Edison's Talking Doll', *Smithsonian Magazine*. Available at: https://www.smithsonianmag.com/smithsonian-institution/epic-failure-thomas-edisons-talking-doll-180955442/ (Accessed: 3 September 2024).

Dia Art Foundation (2023) 'Max Neuhaus, Times Square'. Available at: https://www.diaart.org/exhibition/exhibitions-projects/max-neuhaus-times-square-site (Accessed: 23 September 2024).

Eckel, G. (2001) 'Immersive Audio-Augmented Environments: The LISTEN Project', in *Proceedings Fifth International Conference on Information Visualisation*, IEEE, pp. 571–573. Available at: https://doi.org/10.1109/IV.2001.942112.

Ford (2015) 'How New Ford Technology Can Make Your Car Work Like a Giant Pair of Noise-Cancelling Headphones'. Available at: https://media.ford.com/content/fordmedia/feu/en/news/2015/11/11/how-new-ford-technology-can-make-your-car-work-like—a-giant-pa.html (Accessed: 25 September 2024).

Golden Gate National Parks Conservancy (2007) 'Alcatraz Breaks Out! Major Renovation of World Famous National Park Site', *Golden Gate National Parks Conservancy*. Available at: https://www.parksconservancy.org/news/alcatraz-breaks-out-major-renovation-world-famous-natioenal-park-site (Accessed: 17 September 2024).

Gronow, P. and Saunio, I. (1990) *Äänilevyn historia*. Porvoo: WSOY.

Harman (2024) 'Active Noise Cancellation – The Sound of Silence'. Available at: https://car.harman.com/solutions/car-audio/halosonic/active-noise-cancellation (Accessed: 25 September 2024).

Heide, E. van der (2023) 'Edwin van der Heide – Portfolio'. Available at: https://www.evdh.net/portfolio/EvdH_portfolio.pdf (Accessed: 12 September 2024).

Howard, P. and Templeton, W.B. (1966) *Human Spatial Orientation*. London: Wiley. Available at: https://archive.org/details/trent_0116301611517.

Hyundai (2020) 'Hyundai's World's First Road–Noise Active Noise Control, RANC', Hyundai Motor Group. Available at: https://www.hyundaimotorgroup.com/story/CONT0000000000090151 (Accessed: 25 September 2024).

Kennedy, R. (2006) 'At Museums: Invasion of the Podcasts', *The New York Times*, 19 May. Available at: https://www.nytimes.com/2006/05/19/arts/design/at-museums-invasion-of-the-podcasts.html (Accessed: 3 October 2024).

Kittler, F.A., Winthrop-Young, G. and Wutz, M. (1999) *Gramophone, Film, Typewriter*. Stanford, CA: Stanford University Press (Writing science).

Kotz, L. (2009) 'Max Neuhaus: Sound into Space'. Available at: https://faculty.ucr.edu/~ewkotz/texts/Kotz-2009-Dia-Neuhaus-proof.pdf (Accessed: 23 September 2024).

Kroon, R.W. (2010) *A/V A to Z: An Encyclopedic Dictionary of Media, Entertainment and Other Audiovisual Terms*. McFarland.

Lawton, M., Cunningham, S. and Convery, I. (2020) 'Nature Soundscapes: An Audio Augmented Reality Experience', in *Proceedings of the 15th International Audio Mostly Conference (AM'20). AM'20: Audio Mostly 2020*, Graz Austria: ACM, pp. 85–92. Available at: https://doi.org/10.1145/3411109.3411142.

Majdak, P., Goupell, M.J. and Laback, B. (2010) '3-D Localization of Virtual Sound Sources: Effects of Visual Environment, Pointing Method, and Training', *Attention, Perception, & Psychophysics*, 72 (2), pp. 454–469. Available at: https://doi.org/10.3758/APP.72.2.454.

Miller, G. (2024) 'LINKED'. Available at: https://graememiller.org/project/linked/ (Accessed: 11 November 2024).

Nederlands Instituut voor Beeld & Geluid (2020) 'Draadloze rondleiding in het Amsterdams Stedelijk Museum (1952)'. Available at: https://www.youtube.com/watch?v=QSEqf_mqjCA (Accessed: 2 October 2024).

Niittymäki, H. (2024) 'Email exhange with Matias Harju, 3 October'.

Pettorino, M. (2015) 'The History of Talking Heads: The Trick and the Research', in R. Hoffmann and J. Trouvain (eds) *Proceedings of the First International Workshop of the History of Speech Communication Research. HSCR 2015*, Dresden: TUD Press, pp. 30–41. Available at: https://unora.unior.it/bitstream/11574/160948/1/pettorino_dresden_2015.pdf (Accessed: 19 September 2024).

Pick, H.L., Warren, D.H. and Hay, J.C. (1969) 'Sensory Conflict in Judgments of Spatial Direction', *Perception & Psychophysics*, 6 (4), pp. 203–205. Available at: https://doi.org/10.3758/BF03207017.

Schoenherr, S.E. (1999) 'Sound Recording Research at Bell Labs'. Available at: https://www.aes-media.org/historical/html/recording.technology.history/bell-labs.html (Accessed: 10 September 2024).

Schraffenberger, H. and Heide, E. van der (2014) '*The Real in Augmented Reality*', in xCoAx: Proceedings of the Conference on Computation, Communication, Aesthetics and X. xCoAx 2014, Porto, pp. 64–74.

Self, D. (2024) 'Acoustic Location and Sound Mirrors'. Available at: http://dougla s-self.com/MUSEUM/COMMS/ear/ear.htm (Accessed: 20 September 2024).

Stamp, J. (2023) 'A Partial History of Headphones', *Smithsonian Magazine*. Available at: https://www.smithsonianmag.com/arts-culture/a-partial-histor y-of-headphones-4693742/ (Accessed: 23 September 2024).

Stedelijk Museum Amsterdam (2021) Have You Heard About the Very First Museum Audio Tour?, *Instagram*. Available at: https://www.instagram.com/ stedelijkstudies/p/CT2SHjFIBoS/ (Accessed: 2 October 2024).

Steijn, R. (2023) 'Ode aan de verstilling – Cilia Erens (1946–2023)', *Theaterkrant*. Available at: https://www.theaterkrant.nl/tm-artikel/ode-aan-de-verstilling-cilia -erens-1946-2023/ (Accessed: 9 October 2024).

Thomas A. Edison Inc. (1920) 'The New Edison Advertisement (August 14, 1920)', *The Saturday Evening Post*, 14 August, pp. 164–165.

Thor (2024) 'THOR AVAS – Acoustic Vehicle Alerting System for All Types of Electric Vehicles'. Available at: https://thor-avas.com/legislation/regulatio n-on-avas-requirements/ (Accessed: 25 September 2024).

Waller, S. (2002) 'Psychoacoustic Influences of the Echoing Environments of Pre-historic Art', *Journal of The Acoustical Society of America*, 112, pp. 2284–2284. Available at: https://doi.org/10.1121/1.4779166.

Yu, K. (2022) 'The World's DEEPEST Bin', *Valens Research*. Available at: http s://www.valens-research.com/dynamic-marketing-communique/the-worlds-deepest-bin-find-out-how-this-campaign-encouraged-a-fun-cleaning-a ctivity-in-2009-thursdays-gorillas-of-guerrilla-marketing/ (Accessed: 23 September 2024).

5 Towards mixed realities

Until the 1990s, the merging of virtual audio into the real world was largely restricted to either direct augmentations using hidden loudspeakers or audio guides and audio walks offering linear or manually controlled content through headphones (see Chapter 4). At that time, advances in computing performance were propelling the enthusiastic exploration on virtual realities and 'cyberspace'. In 1992, the term 'augmented reality' was introduced by Caudell and Mizell (1992) referring to their prototype head-mounted display designed at Boeing. It was soon followed by 'augmented audio reality' (Cohen, Aoki and Koizumi 1993) and 'audio augmented reality' (Bederson, 1995) when the authors presented some of the very first practical prototypes of AAR, demonstrating the use of computationally rendered spatial sounds and location-based interactivity.

The journey towards auditory mixed reality (MR) has begun, envisioning a future where computer-mediated virtual sounds seamlessly blend with the real world. While this seamless integration remains an aspirational goal for many practical applications, other characteristic features of AAR have already been successfully leveraged to create engaging and informative experiences. This chapter explores this path through a selection of binaural or headphone-based AAR examples, with the emphasis being on significant milestones in the field as well as narrative applications that are still accessible to the public today. As with the examples from the previous chapter, the examples presented here are referenced throughout the book.

5.1 First AAR prototypes

Since Bell's 'binaural' experiments with two telephone lines (see Chapter 4.1.), telepresence with realistic acoustic representation at the receiver's end has been a one of the core interests in the field of virtual audio. In terms of AAR, an interesting study was conducted by Cohen, Aoki and

DOI: 10.4324/9781003627289-5

Koizumi (1993) at the University of Aizu, Japan. The researchers explored a situation where the operator of a 'telerobot' could hear what the robot hears—with the robot working in a hazardous environment—through a binaural dummy head and headphones. Using 'augmented audio reality' as they called it, this 'true' soundscape would then be overlayed with computer-generated spatialised sounds, such as warnings and communication, as if they coexisted with the real ones.

In addition to augmenting the binaural telepresence feed, Cohen and colleagues also tested to spatially align a virtual ringing sound on a muted desk telephone. This would be one of the earliest reported demonstrations of attaching a binaurally rendered virtual sound on a real-world object (other than loudspeaker).

Whereas in the Aizu experiment, the user was sitting on a chair, Bederson (1995) and his colleagues at the Bell Communications Research in New Jersey, built an influential prototype exploring another important paradigm of AAR: location-based interaction. In their automated museum audio guide, the user was wearing headphones, and when walking up to an exhibit, the system would automatically playback a description of it. When moving away, the description would stop. Bederson emphasised the social aspect this interactive system would offer, enabling friends to navigate the exhibition at their own pace while 'staying in sync' with the same audio content when examining the museum pieces together.

Introducing the term 'audio augmented reality' for the first time, this prototype used an infrared (IR) transmitter at each exhibit and an IR receiver on the headphones. When the receiver picked up the signal from a transmitter, a microprocessor controlled a modified MiniDisc player to choose the corresponding audio description. There was no attempt to spatialise or externalise the sounds as if they were emanating from the environment. However, the audio cues were still spatially and contextually linked to specific real-world objects; to hear the sounds, the participant had to kinaesthetically interact with the objects and the space.

5.2 AAR in exhibitions

Bederson's concept has been highly influential for a multitude of exhibition audio guides. These applications share the primary interaction method of movement and location relative to the exhibits and the exhibition space. The audio-augmented content is usually focused around the exhibits, while the real environment, the venue, is typically ignored. This makes the applications tourable: to accommodate a new exhibition space, merely technical adjustments are required as there is no need to incorporate the local narrative into the experience.

5.2.1 Personalised content

Relatively soon after Bederson's AAR, two ambitious audio guide projects were launched, *LISTEN* in 2001 and *ec(h)o* in 2004. They both utilised binaural spatial audio in 6DoF and experimented with personalised and context-aware audio content. *LISTEN* (e.g., Eckel, 2001; Zimmermann and Lorenz, 2008) was an international project prototyped at the Kunstmuseum Bonn in 2003. The system tracked users with optical IR cameras and adapted to their behaviour, for instance, whether they were sauntering, standing focused at an exhibit, or standing unfocused. By analysing these traits, the system chose personalised audio cues for each visitor. The project also explored the concept of *attractor sounds* (see Chapter 9.12), meant to draw visitors' attention to certain exhibits.

Ec(h)o (Hatala and Wakkary, 2005), in turn, was prototyped at the Canadian Museum of Nature in Ottawa in 2004, and like *LISTEN*, it also provided personalised content based on analysing users' movements. As an experimental user interface, a wooden cube was carried by the users and tracked by computer vision cameras. When arriving at individual artefacts, the cube was used to interact with information about the items in a conversation-like dialogue with the system.

Both systems were well received by participants who found them engaging, informative, and easy to use. One challenge was the need to strike a balance between providing users with a personalised experience and ensuring that the system did not become too distracting. While these two projects demonstrated some of the capabilities of AAR at its best, practical systems in large-scale exhibition have since used less sophisticated but arguably more reliable technology. Also, while highly personalised content would be technically possible to realise, the lack of practical examples since these two projects suggests that productions tend to prioritise other areas of development when using their limited resources.

5.2.2 Large-scale engagement

Perhaps the largest exhibition so far using location-based, interactive audio was *David Bowie Is*. Commissioned by Victoria & Albert Museum and produced by 59 Studio and Real Studios (59 Studio, 2024), it toured at 12 museums around the world from 2013 to 2018. The exhibition could accommodate hundreds of simultaneous visitors. With variations from museum to museum, the exhibition presented around 500 artefacts, music, videos, and information from the life of David Bowie. When a participant moved through the exhibition space, the system played narration and music that changed smoothly, based on proximity with the exhibits. Each user wore headphones and a Sennhei-ser *guidePORT* bodypack, which detected proximity to exhibits equip-ped with RF transmitters and played back the corresponding content (Sennheiser, 2013). Whenever a video was shown, the audio on the headphones was synced with the visuals. Despite the audio being head-locked and the users' absolute location not being tracked, many found the experience extremely immersive and captivating. This serves as an excellent example of how innovative design and high production values can create strong engagement, even with limited sensory immersion.

One can consider the *Dimensions of Sound* exhibition at the *House of Music* in Budapest, Hungary, as an updated version of *David Bowie Is*. Opened in 2022, this large permanent exhibition is realised with the *usomo* system, capable of 6DoF using UWB (ultra wide-band) tracking and spatial audio. The exhibition is based on interactive engagement with the exhibits with dynamic, location-triggered and video-synced audio content. Head tracking enables sounds to be attached to their corresponding physical objects, enhancing immersion and clarity. Designed to accommodate dozens of simultaneous visitors, the experi-ence does not strive for realistic auralisations, but relies on rich content and interactivity. Like *David Bowie Is*, this exhibition showcases AAR in a museum context through a robust, large-scale system capable of effi-ciently handling high visitor volumes.

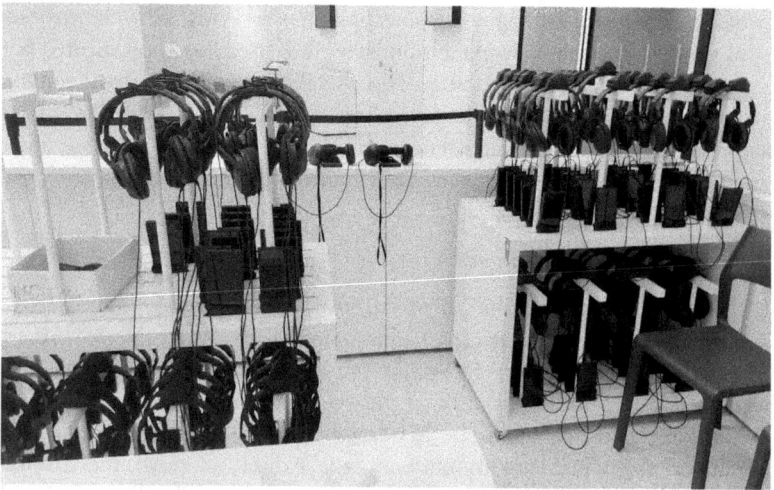

Figure 5.1 Headphones and mobile devices equipped with the *usomo* system waiting for visitors at the *Dimensions of Sound* exhibition in Budapest
Source: Photograph by the author

5.3 Reinforced environments

The unique nature of AR and AAR is their capability to reveal new meanings in the reality. Whereas the aforementioned examples focused on adding virtual sounds mainly on real-world objects, it is also possible to *reinforce* the whole environment (Azuma, 2015), discussed more in Chapter 8.5. The audio content in the following examples is rooted in the place and its people. Such experiences are obviously site-specific and cannot be transferred to elsewhere.

5.3.1 Augmented office

Audio Aura was an early AAR research project at the Xerox PARC in California (Mynatt et al., 1998). The goal was to use virtual background sounds to provide 'serendipitous' information about, for example, colleagues' whereabouts in the workplace. When the user dropped by at someone's office—only to find the room empty—the system would play an audio cue to the user's wireless headphones and inform whether the colleague was to be expected coming back or not. RF proximity sensors were used to track each employees approximate location. In addition to this location-based, socially meaningful information, the system worked as a ubiquitous audio display by informing the user of the amount of received emails. All audio cues and sonifications were nature sounds such as seagulls and waves to provide 'peripheral' information rather than exact, 'a sense akin to seeing [the colleague's] light on and their briefcase against the desk' (Mynatt et al., 1998).

Even though test users reacted positively, systems like *Audio Aura* have not gained traction in workplaces. While there were no technical barriers to adoption, the concept may have simply been ahead of its time. Additionally, privacy concerns could have been a potential obstacle, along with the possibility that constantly hearing ambient sounds as reminders might become stressful over time, regardless of how serendipitous they were.

5.3.2 Time travelling

In narrative AAR, virtual background ambiences have of course been used a lot. One quite recent example where the real-world place is enhanced with imaginary atmospheric sounds and dialogue is *Sonic Traces Heldenplatz* in Vienna. Launched in 2020 and developed by scopeaudio and hoerwinkel (Aichinger, 2024), the experience is set at Heldenplatz, a central venue in the history of Austria. Participants wear a pair of head-tracking *Bose AR* headphones and carry a mobile phone with a simple graphical user interface (GUI), which allows them to switch between two time layers—1848 and

2084—as they stroll across the square. In each layer, the user finds acousmatic characters discussing the pressing issues of their time, while the soundscape portrays the period's atmosphere with horse carriages and flying cars. Interestingly, when switching layers, the characters continue debating the same themes, though from the distinct perspectives of their respective eras.

The immaterial audio layer brings back history also to the *Maison Gainsbourg* in Paris. It is nowadays the home museum of the French musician Serge Gainsbourg and his family. Opened in 2023 to the public, it presents the house as it was left in 1991 when the musician died. The museum is presented through a 30-minute, pre-booked immersive audio guide using the *usomo* 6DoF system. The audio content is created by The Soundwalk Collective with Charlotte Gainsbourg, the daughter of Serge Gainsbourg. During the tour, the visitor hears Charlotte reminiscing about her and her father's life in the house. A unique feature about the tour is the use of original recordings of the family living their everyday life in the house. When these recordings are spatially attached to the same locations where they were originally recorded—for instance, Serge teaching Charlotte to play piano heard from the direction of the grand piano in the living room—the temporal layers align in a powerful way.

Figure 5.2 Visitors waiting for their audio tour to start at *Maison Gainsbourg* in Paris
Source: Photograph by the author

5.4 Reskinned environments

Applications set in a 'reskinned environment' are either set in an artificial environment that is custom-built for the experience, or in a real place that is refurbished for the use (Azuma, 2015), as discussed further in Chapter 8.5. They can usually be transported to other places because, like exhibition AAR, reskinning pays little or no attention to the real world outside the 'magic circle' of the experience.

5.4.1 Poetry of objects

A good example of reskinning and a custom-built set is *The Sound of Things*, a sound installation by Holger Förterer (2013). The installation, set in a dimly lit room, comprises a wooden table with a table lamp casting light on ten items carefully scattered across its surface. Two simultaneous listeners wear wireless headphones tracked with infrared cameras in 6DoF. When the listener crouches closer to the objects, virtual sounds can be heard coming from them: a wine glass emits sounds of celebration, a memory pad produces sounds of scribbling, the desk lamp hums softly, the cigarette sputters... The lifeless items seem to whisper memories, and though it is hard to piece together a coherent story from them, the listener starts to wonder whose desk and belongings these are, what happened to the owner, who were the people socialising around the wine glass, what is written on the memory pad.

The experience can be perceived as an acoustic, three-dimensional poem or a scene from a mystery tale. *The Sound of Things* serves as an

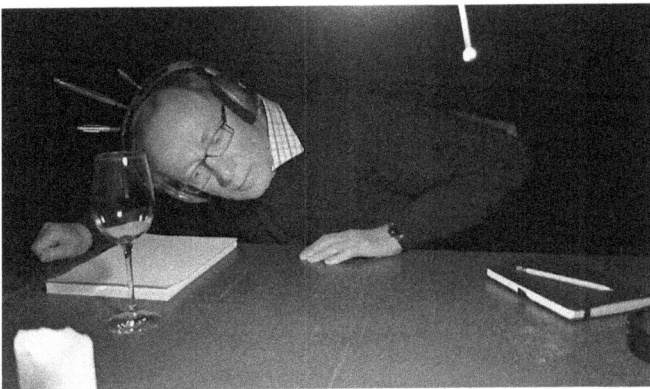

Figure 5.3 A participant in *The Sound of Things* installation
Source: Still capture from video by Holger Förterer

example of using 6DoF binaural AAR efficiently as a medium for impli-
cit, fragmented narratives. The fact that the environment around the
user fades into darkness and gets ignored further adds to framing the
piece as symbolic rather than real.

5.4.2 Narrative reskinning

Darkness is the central feature of the productions by the British company
DARKFIELD. They make audio-based experiences in total darkness, built
inside shipping containers. While containers provide total control over the
narrative environment, they also enable global touring and establish clear
boundaries for the creative process. DARKFIELD draws from the immersive
theatre tradition, and their first show *Séance* was premiered in 2018. It serves
as a prime example of their approach, where the audience experiences a 20–
30 minute linear narrative with binaural sound, tightly connected to both the
fabricated environment and the real patrons participating in the show. The
fact that *Séance* takes place in darkness makes acousmatic characters and
events feel plausible while fostering imagination.

Figure 5.4 Two DARKFIELD shows, *Séance* and *Arcade*, on tour in June 2024 in
Norwich, UK
Source: Photograph by the author

Whereas *The Sound of Things* augments staged objects on a table and *Séance* enhances a fully prepared shipping container, *Horror-Fi Me* by Laurence Cliffe (2021; 2024) reskins almost any indoor environment with a horror movie soundscape. This iPhone app allows users to attach virtual sounds to objects in their home: for example, windows are augmented with rain and storm effects, doors with occasional knocking, and faucets with dripping water. Using spatialised sounds with 6DoF, these augmentations envelop the user and can emotionally transform a safe environment such as a home into a space of dread. As a context-aware application, *Horror-Fi Me* is also a compelling example of both transferable and environment-driven AAR, discussed further in Chapter 8.4.

As the last example of reskinning, *The Reign Union* (*Valtakunnan liitto*) uses a real-world site as the foundation of the story but reskins it throughout the experience. It is a biofictional AAR story created as part of the artistic research project 'Full-AAR' at WHS in Helsinki, Finland. Open to the public since May 2025 and located in a gallery space that used to accommodate a laundry in the 1950s, the story revolves around the connections between the founder of the laundry and a subversive, underground Nazi group during World War II. Two simultaneous users navigating the space share the same story, but hear different narratives, leading to questions on who to trust, and whether to trust what one hears.

The gallery natively serves as a transitional space, positioned between two exhibitions, with a few crates and pieces of equipment scattered around. This makes it an ideal platform for reskinning—transforming it from its current state into any other setting, such as a laundry, railway station, or garden, both contextually and through virtual acoustics.

Full-AAR

The Full-AAR project, launched in 2021 and run by WHS, a Finnish contemporary circus and visual theatre group, explores the narrative possibilities and technical approaches of headphone-based AAR using 6DoF. The author of this book works as the project lead and sound designer on the project. As part of Full-AAR, a technical platform was developed to track multiple participants indoors using depth cameras, body-tracking algorithms, and custom headphones. *The Reign Union* story was produced to test and demonstrate the platform and experiment with several narrative techniques characteristic of AAR, some of which are discussed in Chapter 9.

5.5 Navigation and situational awareness

AAR has huge potential in providing navigational aid without the restrictions of obtrusive visual screens and enforcing situational awareness through omnidirectional hearing. This was recognised early on, and for instance, since the 1970s, audio-based assistance systems for people with visual impairment have been developed. The first systems used fixed installations of infrared transmitters on objects of interest, sending audio marks to the user (Schmalstieg and Hollerer, 2016). Later, satellite positioning and spatialised, text-to-speech voice attached on waypoints were prototyped, for example, in *Personal Guidance System* at the University of California Santa Barbara (Loomis, Golledge and Klatzky, 1998) and *NAVIG* by a France-based consortium 2008–2014. The NAVIG system was rather sophisticated: besides outdoor environments, it worked indoors and used head-mounted cameras to recognise objects and features such as stairways, augmenting them with spatialised voices and various sonifications (Katz et al., 2012). Nowadays, as discussed in Chapter 2.8, there are a number of mobile phone applications that use computer vision and AI to interpret the surroundings, although they do not usually support spatial audio.

Yet, navigation using spatial audio cues would be useful for sighted people, too. This concept has been explored in, for instance, *Guided by Music*, a study by Albrecht, Väänänen, and Lokki (2016). Their prototype directed walkers and cyclists to their destinations by spatialising music to the waypoints along the route, thus avoiding looking at a screen for directions. However, the lack of head-tracking headphones—until recently—has probably prevented the emergence of applications using spatialised audio beacons for navigation. Therefore, the head-locked turn-by-turn voice prompts of *Google Maps* and other apps remain the primary navigation aid when the visual domain is not available or preferred.

One industry that has put significant research efforts into virtual binaural audio since the 1990s is aviation. Its motivation has been to develop collision avoidance systems within general aviation as well as handle threat-related tasks and voice separation in military cockpits (Simpson et al., 2005). Multiple flight simulator studies have shown that 3D audio displays do help pilots to react faster to threats and events around the aircraft while reducing cognitive load from the visual modality (Veltman, Oving and Bronkhorst, 2004).

The first actual 3D audio systems integrated in pilots' helmets with head tracking were manufactured by the Danish company Terma and installed on F-16 fighter jets of the Royal Danish Air Force in the 2010s

(Kucinski, 2018). These and other similar systems have two main features. Firstly, they spatially separate radio channels so that the pilot stays aware of all the communication but can focus on the most important ones; the voices of the crew members in the intercom are also spatialised, making communication more natural and efficient (Szondy, 2013; Terma, 2024). Secondly, the threats are indicated by sounds spatialised to the actual directions, increasing situational awareness and reducing reaction time significantly compared to traditional panel-mounted displays (Kucinski, 2018; Evans, 2024).

5.6 Geolocated audio

Accurate satellite navigation revolutionised many industries and fields, and gave birth to completely new services. In May 2000, the US government disabled the deliberate GPS error for civilians, improving the accuracy of the GPS tracking from 45–100 metres to about 5–15 metres (NOAA, 2018). This propelled the development of geolocated services and applications. It took, however, a bit less than a decade before GPS receivers were integrated in mobile phones with a fast internet connection and easy-to-use graphical operating system. Importantly, the APIs of map services such as *MapQuest, OpenStreetMap*, and *Google Maps*, were opened for developers. When iPhone 3G was launched in 2008, these were all available for developers. Later, three other satellite navigation services became available to the public, commonly called *global navigation satellite systems* or *GNSS*, and the accuracy of the positioning further increased significantly.

This eventually made *geolocated AAR* possible, allowing anyone to make, share and enjoy content embedded in the environment. In geolocated AAR, a dedicated software platform is used, where sound files are attached to 'geofenced' areas on a map. When the user, carrying a mobile phone and headphones, enters one of the areas, the assigned sounds are played. Geolocated applications fall roughly into two categories: 1) multi-authored, social apps where users can record their own sounds and place them on a map to be later discovered by other users, and 2) audio walks or audiotrails, which are usually site-specific and authored by a single creator.

5.6.1 Multi-authored apps

One of the first GPS-based AAR experiments was *Hear&There*, prototyped at the MIT Media Lab (Rozier, Karahalios and Donath, 2000). Like Bederson's concept, it emphasised the social aspect, exploring a platform where people could interact by leaving personal auditory 'imprints' within

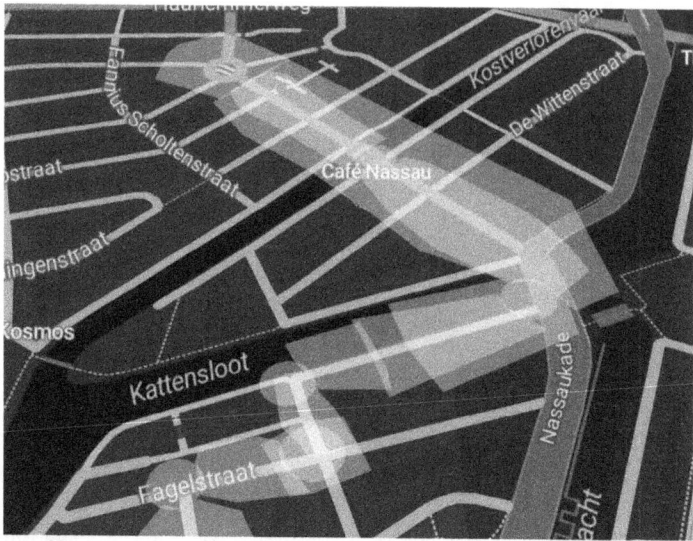

Figure 5.5 A geolocated audio walk made and shared on *Echoes* with geofenced trigger areas
Source: A screenshot from the Echoes iOS app

the campus premises and discovering those left by others. By moving beyond proximity sensors, the use of GPS enabled completely arbitrary positioning of sounds anywhere the user desired, unlocking significant possibilities for future location-based audio applications. *Hear&There* also tracked users' head orientation, enabling sound spatialisation in XY-Yaw degrees (see Chapter 2.12), a feature that remains rare in geolocated experiences today.

Later, open platforms for geolocated sound sharing became available for public, *Roundware*, released in 2008 and still available, being one of the first (Burgund, 2024). Since, there have been many other platforms, most of which have, however, been short-lived. This kind of format may suffer from the lack of cohesive content and interaction with other people, making the threshold to actually go to places to hear the messages and sounds rather high.

5.6.2 Single-authored apps

Single-authored geolocated experiences, on the other hand, are quite popular. These are typically audio walks or audiotrails, supported by a variety of mobile applications and platforms enabling the authoring and sharing of map-based audio content with automatic playback triggered by the user's geolocation. In the 2010s, several platforms were released, including *NoTours* in 2010, *SonicMaps*, 2012, *Echoes*, 2013, *Soundtrackcity*, 2013, and *U-Gruve* in 2016. In the authoring stage, these applications offer the user a set of predefined tools for drawing 'geofenced' areas on a map layout and assigning audio clips for them. In runtime, the sounds are controlled based on a participant's movements in and out of the areas, and sometimes within them.

Geolocated audio is used among independent artists and houses of culture for sharing site-specific memories, drama, poetry, and music. Some platform providers, such as *Echoes*, also create commissioned experiences and bespoke apps based on their technical framework (Kopeček, 2024). Alongside audiotrails, interactive tourist audio guides have emerged, using the same principles.

Further, besides walking, nothing prevents creating content for cars, bicycles, watercraft, or any other method of movement. More recently, *The Sonic Bike* project used bicycle-mounted speakers for geolocated music and sonic art (Bicrophonic Research Institute, 2024). *Audio Nomad* (Woo et al., 2006) augmented passing seascape with surround loudspeakers. *Autio*, co-founded by actor Kevin Costner, offers geolocated stories with high production value mainly targeted at people traveling by car. While satellite navigation limits the experiences set outdoors, some platforms offer coarse indoor tracking, such as *Echoes* that supports iBeacons for indoor proximity interaction, based on Bluetooth Low Energy (BLE) technology.

5.6.3 Geolocated experiences

One intriguing example of the use of the geolocated audio platforms is *Hyperkuulo* ('Hyper hearing'), an interactive walk built on the *NoTours* platform. It was launched in April 2013 in Tampere, Finland, by Ari Koivumäki (2018) with students of the Tampere University of Applied Sciences. The walk featured a narrative where users participate in a scientific experiment with an implant that enhances hearing. Headphones were integral to the story, framing all sounds as part of the experiment and partially excused the use of head-locked audio. Set in a park, users completed tasks that engaged them with their surroundings, such as walking around a fountain counterclockwise while humming 'omm', and making noises on an outdoor bandstand to create reverberations. Due to the enhanced hearing sensitivity, the user could hear, for instance, grass grow, people bathe inside a distant sauna, and sounds of the past and future. The narrative progressed automatically as the user reached designated locations, but since the system lacked sensors to verify their actions, it relied solely on the player's willingness to follow the instructions while maintaining an element of the illusion of agency (see Fendt et al., 2012).

Whereas *Hyperkuulo* was site-specific, *The Planets* (Sofilab, 2023) is an example of a transferable geolocated application. It is available in multiple urban parks around the world, authored remotely using *Google Maps* (Nitschke, 2022). The listener, equipped with a mobile phone and headphones, assumes the role of a spaceship pilot and navigates from planet to planet by walking within the park, guided by a binaurally rendered sound signal. Each planet, or designated area in the park, features the corresponding movement from Gustav Holst's orchestral suite *The Planets*, played back as spatialised acousmatic instrument groups. While *The Planets* is set in a real-world environment and its primary interaction mode is physical movement within that space, the content itself lacks a direct con-textual relationship with the surroundings. This forces listeners to actively interpret and construct their own connections between the space theme, the music, and the urban park.

5.7 AAR games

AAR games are a niche genre, but nevertheless, remain an interesting platform for studying interactivity and immersion in an audio augmented environment. They differ from audio games by bringing gameplay and stories physically to the real world, letting the player interact with the game kinaesthetically, and interact with the reality through the game.

5.7.1 Environmental relationship

Some of the earliest AAR games used the real environment merely as a platform for walking and thus interacting with the game, not unlike *The Planets*. For instance, *Guided by Voices* (Lyons, Gandy and Star-ner, 2000), a prototype developed at the Georgia Institute of Technol-ogy, USA, immersed the player into a fantasy world through virtual sounds without any narrative relationship with the real environment. Gameplay was based on making decisions in the acoustic fantasy world by physically moving between several locations within the real space, which could be an office, yard, or any other place. Despite the obvious incongruence between the real and virtual worlds, the movement reportedly aided in players' immersion in the game. The game also fostered social interaction as multiple people could play it simultaneously in the same space.

A more recent AAR game prototype, *Sound PacMan* (Chatzidimitris, Gavalas and Michael, 2016), was, on the other hand, completely based on the user's environment. The gameplay followed the idea of the original *Pac-Man*, but transformed the user's local street network into the game's maze; the player's objective was to traverse along the real-world streets and eat 'cookies' while avoiding the pursuing ghosts that were also following along the same streets. Both cookies and ghosts were denoted by virtual sounds. The ubiquitous and adaptive nature of the game fundamentally leveraged the unique nature of AAR as a layer to reality, even if the contextual connection between the game characters and the local streets may have remained vague.

In contrast, the narrative connection with the environment was strong with *Audio Legends* (Rovithis et al., 2019; Moustakas et al., 2020), playable in 2018–2019. It was a site-specific outdoor game created by researchers at the Ionian University in Greece, and based on the folklore of the island of Corfu. The player's task was to protect the island by escorting a sailor carrying wheat to the town's entrance, locating and fighting 'the monsters of plague', and blocking incoming cannonballs (Rovithis, 2024). The real environment was utilised by embedding gameplay elements in the appropriate real-world surroundings, such as enemy gunships on the sea. The game also explored the use of gestures as an interaction mechanism within AAR: for instance, the user had to move the mobile device as if it were a shield to block the acousmatic cannonballs. The authentic environment and the contextual links between it and the story seem to have played a significant role in making the game feel immersive to the players (Rovithis et al., 2019).

Zombies, Run! takes an interesting approach to the user's environment. Released in 2012, this AAR application immerses the user in a zombie narrative set in their own neighbourhood. As 'Runner Five', one of the few survivors, the user must run and evade zombies while completing game-like tasks to save the community. The user can listen to their own music playlist on a mobile device, which is, every now and then, interrupted by a 'radio transmission' where the protagonists talk to the user. The application also features *Radio Mode* that augments the listener's music playlist with virtual radio broadcasts, creating an interesting layered reality discussed in Chapter 3.1.

The application does not attempt to embed virtual sounds in the environment other than with some general ambiences. In fact, location tracking is not utilised, meaning the system has no awareness of the user's whereabouts or the nature of their surroundings. Instead, the gameplay relies on the user's pace, with elements like randomly occurring chases where zombies' breathing and drooling can be heard behind

Figure 5.6 A player of *Audio Legends* blocking an incoming cannon ball with the
          mobile device
Source: Photograph courtesy of Emmanouel Rovithis and Nikolaos Moustakas

the user, prompting them to speed up for a minute to evade capture.
Without head tracking, the app assumes users predominantly look forward while running. Despite not knowing the user's surroundings, the
dialogue references generic landmarks and directions ('hospital', 'old
mill', 'east of the trees'), creating an illusion that the story world aligns
with the real one.

5.7.2 Collaboration

The aforementioned AAR games were designed for solo play, although
multiple people could play them at once, sharing experiences in the same
space. *Please Confirm You Are Not a Robot* was, however, an experimental game as part of a study that explored participatory performance
and collaborative storytelling in AAR (Nagele et al., 2021). Realised in
2019 in London, UK, each performance accommodated four participants

wearing *Bose Frames* smart glasses with head tracking and open-ear headphones. The participants were engaged in a series of games where they interacted with spatial audio cues as well as each other. As the game progressed, the virtual narrator 'Pi' revealed personal feelings and fears to each player individually, and eventually asked one participant to speak on their behalf to the rest of the group, thus testing the trust in the narrator and the game. The game provided players with asymmetric information, making participants hear different elements of the story, similar to *The Reign Union*. This prompted group coordination through verbal and non-verbal communication. Furthermore, the use of spatial audio cues prompted participants to engage their bodies in the performance, leading to a feeling of presence and immersion.

5.8 Conclusion

The examples examined in this chapter highlight some of the potential of binaural AAR, particularly in their relationship with reality. While many fascinating projects have been omitted to maintain focus, and developments in virtual audio research and technology are reserved for other discussions, this chapter aims to provide a general overview of achievements in this field. One significant challenge in learning about AAR experiences is the lack of a standardised vocabulary. This results in content creators, museums, institutions, and product manufacturers using varied terms to describe features that could readily be classified as AAR. At times, exhibitions with compelling AAR elements may go unpromoted because curators or press officers fail to recognise them as noteworthy. Nevertheless, as long as innovative content continues to be created, technology advances to support those ideas, and these creations are effectively communicated to audiences, we can set aside terminology and focus on fully immersing ourselves in the world of AAR.

References

59 Studio (2024) 'David Bowie Is', *59 Studio*. Available at: https://59.studio/project/david-bowie-is/ (Accessed: 24 November 2024).

Aichinger, T. (2024) 'Conversation with Matias Harju, March 6'.

Albrecht, R., Väänänen, R. and Lokki, T. (2016) 'Guided by Music: Pedestrian and Cyclist Navigation with Route and Beacon Guidance', *Personal and Ubiquitous Computing*, 20 (1), pp. 121–145. Available at: https://doi.org/10.1007/s00779-016-0906-z.

Azuma, R. (2015) 'Location-Based Mixed and Augmented Reality Storytelling', in B. Woodword (ed.), *Fundamentals of Wearable Computers and Augmented Reality*, 2nd edn. CRC Press, pp. 259–276.

Bederson, B.B. (1995) 'Audio Augmented Reality: A Prototype Automated Tour Guide', in *Conference Companion on Human Factors in COMPUTING Systems – CHI '95*. Denver, Colorado, United States: ACM Press, pp. 210–211. Available at: https://doi.org/10.1145/223355.223526.

Bicrophonic Research Institute (2024) 'Bicrophonic Research Institute – Makes Music and Audio Landscapes to be Triggered by You the Cyclist'. Available at: https://sonicbikes.net/ (Accessed: 30 October 2024).

Burgund, H. (2024) 'Email with Matias Harju, November 18'.

Caudell, T. and Mizell, D. (1992) 'Augmented Reality: An Application of Heads-Up Display Technology to Manual Manufacturing Processes', in *Proceedings of the Twenty-Fifth Hawaii International Conference on System Sciences*, pp. 659–669 vol. 2. Available at: https://doi.org/10.1109/HICSS.1992.183317.

Chatzidimitris, T., Gavalas, D. and Michael, D. (2016) '*SoundPacman: Audio Augmented Reality in Location-Based Games*', in 18th Mediterranean Electrotechnical Conference (MELECON). Lemesos, Cyprus: IEEE, pp. 1–6. Available at: https://doi.org/10.1109/MELCON.2016.7495414.

Cliffe, L. (2021) 'Horror-Fi Me'. Available at: https://lozcliffe.com/horror-fi-me/ (Accessed: 9 December 2024).

Cliffe, L. (2024) 'Into the Here and Now: Explorations within a New Acoustic Virtual Reality', *Leonardo* [Preprint]. Available at: https://nottingham-repository.worktribe.com/output/42480431/into-the-here-and-now-explorations-within-a-new-acoustic-virtual-reality (Accessed: 2 December 2024).

Cohen, M., Aoki, S. and Koizumi, N. (1993) 'Augmented Audio Reality: Telepresence/VR Hybrid Acoustic Environments', in *Proceedings of 1993 2nd IEEE International Workshop on Robot and Human Communication*, pp. 361–364. Available at: https://doi.org/10.1109/ROMAN.1993.367692.

Eckel, G. (2001) 'Immersive Audio-Augmented Environments: The LISTEN Project', in *Proceedings Fifth International Conference on Information Visualisation*, pp. 571–573. Available at: https://doi.org/10.1109/IV.2001.942112.

Evans, S. (2024) 'PTC, BAE Team to Create Augmented Reality Fighter Pilot Helmet'. Available at: https://www.iotworldtoday.com/metaverse/ptc-bae-team-to-create-augmented-reality-fighter-pilot-helmet- (Accessed: 10 October 2024).

Fendt, M.W. et al. (2012) 'Achieving the Illusion of Agency', in D. Oyarzun et al. (eds) *Interactive Storytelling*. Berlin, Heidelberg: Springer, pp. 114–125. Available at: https://doi.org/10.1007/978-3-642-34851-8_11.

Förterer, H. (2013) 'The Sound of Things'. Available at: https://www.foerterer.com/sound-of-things.html (Accessed: 3 December 2024).

Hatala, M. and Wakkary, R. (2005) 'Ontology-Based User Modeling in an Augmented Audio Reality System for Museums', *User Modeling and User-Adapted Interaction*, 15 (3), pp. 339–380. Available at: https://doi.org/10.1007/s11257-005-2304-5.

Katz, B.F.G. et al. (2012) 'NAVIG: Augmented Reality Guidance System for the Visually Impaired', *Virtual Reality*, 16 (4), pp. 253–269. Available at: https://doi.org/10.1007/s10055-012-0213-6.

Koivumäki, A. (2018) *Maiseman äänittäminen: Äänimaisematutkimus ääni-suunnittelun tukena*. Helsinki: Aalto-yliopiston taiteiden ja suunnittelun korkeakoulu (Aalto-yliopiston julkaisusarja).

Kopeček, J. (2024) 'Conversation with Matias Harju, October 21'.

Kucinski, W. (2018) 'A-10C Pilots Will Get 3D-Audio to Increase Situational Awareness'. Available at: https://www.sae.org/site/news/2018/11/a-10c-pilots-will-get-3d-audio-to-increase-situational-awareness (Accessed: 10 October 2024).

Loomis, J.M., Golledge, R.G. and Klatzky, R.L. (1998) 'Navigation System for the Blind: Auditory Display Modes and Guidance', *Presence: Teleoperators and Virtual Environments*, 7 (2), pp. 193–203. Available at: https://doi.org/10.1162/105474698565677.

Lyons, K., Gandy, M. and Starner, T. (2000) 'Guided by Voices: An Audio Augmented Reality System', in *Proceedings of the International Conference on Auditory Display April, 2000*. Georgia Institute of Technology, Atlanta, Georgia, USA: International Community for Auditory Display. Available at: https://smartech.gatech.edu/handle/1853/50672 (Accessed: 5 November 2018).

Moustakas, N. et al. (2020) 'Adaptive Audio Mixing for Enhancing Immersion in Augmented Reality Audio Games', in *Companion Publication of the ICMI 2020 International Conference on Multimodal Interaction*. Virtual Event Netherlands: ACM, pp. 220–227. Available at: https://doi.org/10.1145/3395035.3425325.

Mynatt, E.D. et al. (1998) 'Designing Audio Aura', in *Proceedings of the SIGCHI Conference on Human Factors in Computing Systems – CHI '98*. SIGCHI, Los Angeles, California, United States: ACM Press, pp. 566–573. Available at: https://doi.org/10.1145/274644.274720.

Nagele, A.N. et al. (2021) 'Interactive Audio Augmented Reality in Participatory Performance', *Frontiers in Virtual Reality*, 1, p. 610320. Available at: https://doi.org/10.3389/frvir.2020.610320.

Nitschke, M. (2022) 'Conversation with Matias Harju, September 14'.

NOAA (2018) 'GPS.gov: Data From the First Week Without Selective Availability'. Available at: https://www.gps.gov/systems/gps/modernization/sa/data/ (Accessed: 25 November 2024).

Rovithis, E. (2024) 'Audio Legends – SonicManos'. Available at: https://sonicmanos.com/index.php?id=audio-legends (Accessed: 15 October 2024).

Rovithis, E. et al. (2019) 'Audio Legends: Investigating Sonic Interaction in an Augmented Reality Audio Game', *Multimodal Technologies and Interaction*, 3 (4), p. 73. Available at: https://doi.org/10.3390/mti3040073.

Rozier, J., Karahalios, K. and Donath, J. (2000) 'Hear&There: An Augmented Reality System of Linked Audio', in *Proceedings of the International Conference on Auditory Display April, 2000*. Georgia Institute of Technology, Atlanta, Georgia, USA: International Community for Auditory Display.

Schmalstieg, D. and Hollerer, T. (2016) *Augmented Reality: Principles and Practice*. Addison-Wesley Professional.

Sennheiser (2013) '"David Bowie Is": Sennheiser Helps the V&A Bring Together Sound and Vision'. Available at: https://web.archive.org/web/20160112020455/

https://en-de.sennheiser.com/news-david-bowie-is-sennheiser-helps-the-va-bring-together-sound-and-vision- (Accessed: 13 August 2024).

Simpson, B. et al. (2005) 'Spatial Audio Displays for Improving Safety and Enhancing Situation Awareness in General Aviation Environments', in *New Directions for Improving Audio Effectiveness. Meeting Proceedings RTO-MP-HFM-123, Paper 26*, Neuilly-sur-Seine, France, pp. 26. 21–26. 16. Available at: https://apps.dtic.mil/sti/citations/ADA454658 (Accessed: 21 September 2024).

Sofilab (2023) 'The Planets, The Planets'. Available at: https://the-planets.app/ (Accessed: 31 October 2024).

Szondy, D. (2013) 'Raytheon Developing 3D Hearing for Pilots', *New Atlas*. Available at: https://newatlas.com/raytheon-3d-audio/26839/ (Accessed: 10 October 2024).

Terma (2024) '3D-Audio and Active Noise Reduction - Rotary Wing'. Available at: https://www.terma.com/media/a52digfm/3d-audio_rotary_a4.pdf (Accessed: 10 October 2024).

Veltman, J.A., Oving, A.B. and Bronkhorst, A.W. (2004) '3-D Audio in the Fighter Cockpit Improves Task Performance', *The International Journal of Aviation Psychology*, 14 (3), pp. 239–256.

Woo, D.et al. (2006) 'Audio Nomad', in *Proceedings of the Institute of Navigation – 19th International Technical Meeting of the Satellite Division, ION GNSS 2006*, 5, pp. 3117–3123.

Zimmermann, A. and Lorenz, A. (2008) 'LISTEN: A User-Adaptive Audio-Augmented Museum Guide', *User Modeling and User-Adapted Interaction*, 18 (5), pp. 389–416. Available at: https://doi.org/10.1007/s11257-008-9049-x.

6   Spatial hearing and virtual audio

The goal of many AAR applications is to create an illusion of artificial sounds seamlessly coexisting with reality. This chapter approaches this challenge in binaural systems, guiding the reader through the basics of spatial hearing and some typical audio paradigms applicable for AAR. It explores the spatialisation of virtual sounds through key processes such as auralisation, room impulse response acquisition, and binaural decoding, providing insights into achieving externalised, realistic soundscapes that blend with the real environment.

6.1 Sound and auditory events

Sound is produced when a source, like a vibrating guitar string, creates pressure waves that travel through a medium, such as air, water, or solid material. When these waves reach the ear, they move through the ear canal and make the eardrum vibrate. The tiny bones in the middle ear amplify these vibrations and pass them to the cochlea. There, movement of fluid activates hair cells, thus generating electrical signals, which travel via the auditory nerve to the brain and are interpreted as sound (Hudspeth, 1997).

To distinguish the actual physical sound from the perceived sound, we can use Jens Blauert's (1997) concepts of *sound event* and *auditory event*. Sound event is the real-world physical occurrence that generates the sound waves. Auditory event, in turn, refers to the subjective perception of the sound, that is, the cognitive interpretation of the sound waves by the auditory system. In the classic thought experiment, if a tree falls in a forest and no one is around to hear it, there is a sound event but no auditory event.

The auditory event entails not just the original sound, but also its direction, distance, and how the acoustics and obstacles change the sound along its path. When spatial hearing works perfectly, the auditory event aligns with the sound event.

DOI: 10.4324/9781003627289-6

Auditory event (perception)
Sound event (real sound)

Figure 6.1 Auditory event (perception) aligned with the sound event (real-world sound occurrence)

In virtual and augmented realities, the concept of auditory event is important. When working with AAR, we want to trick the brain to believe that a virtual sound is actually emanating from the real environment, possibly from a real-world object. To make that illusion work, we need to create an artificial auditory event that imitates the perceptual attributes of a similar, real-world sound occurrence. Multiple factors need to be taken into account, such as the correct direction and distance of the sound, the assumed sound wave propagation through the environment with any occlusion, diffraction, and reverberation along the path, and finally the effect of head, torso, and outer ears. In an AAR system, if the virtual audio system works perfectly, the auditory event aligns with the intended virtual sound event, which is often attached to a real-world object. However, in practice, that is not always the case, and the sound appears in a wrong place—for instance, too close to the listener, or too high.

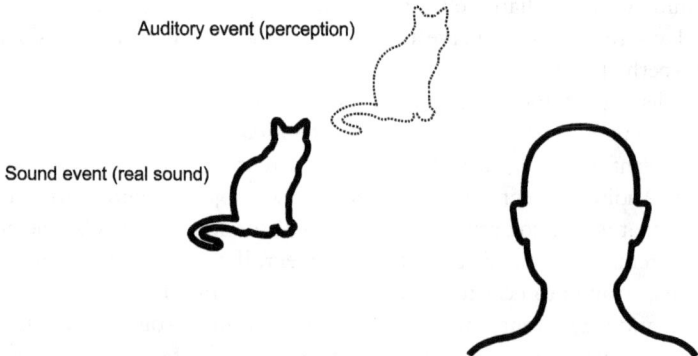

Auditory event (perception)

Sound event (real sound)

Figure 6.2 Auditory event misaligned with the virtual sound (attached to a real-world object) due to imperfect virtual audio rendering

If the artificial sound is played through a hidden loudspeaker from the location where the real-world sound would be ('direct augmentation'), we obviously do not need to worry about the sound spatialisation and propagation effects as they happen naturally—although the directivity of the loudspeaker might be different from the real sound source, thus causing acoustic differences. In *virtual audio* technologies that use *binaural* reproduction, however, it is important to understand which perceptual cues are needed to trick the auditory system and how those illusions can be made. Many of the same considerations are also elementary in the audio production stage even if the delivery format was a surround loudspeaker system.

An auditory event is not always caused by a sound event and sound waves, tinnitus being a prime example (Blauert, 1997). It is possible to generate virtual sounds by stimulating the auditory nerve or brainstem directly (see Veronese et al., 2023), or even stimulating the auditory cortex in the brain (see Leaver, 2024), albeit such techniques are still far from being applicable in practical AAR applications. Further, even if such technologies were available for the virtual audio community, it would not free us from simulating the same acoustic phenomena described above and discussed in more detail in this chapter.

6.2 Spatial hearing

Humans' ability to localise sounds is mainly based on comparing acoustic information between two ears, called binaural hearing. The two key mechanisms of binaural hearing are interaural time difference (ITD) and interaural loudness difference (ILD) (Blauert, 1997). However, there are other psychoacoustic cues in play, such as caused by head, torso, and outer ears, visual cues, recognising the acoustic environment, room congruence, and motion within the space (Brandenburg et al., 2023).

An ITD is caused when the sound source is, for example, on the left of the listener in a *horizontal plane*; the sound waves will first reach the left ear, and after a small delay, the right ear will register the sound. The maximum ITD, around 650 microseconds (in air at room temperature) occurs when the sound is 90 degrees to the side (Wenzel, Begault and Godfroy-Cooper, 2018). However, when the sound originates from the *median plane*—that is, directly in front, behind, above, or below—both ears will receive the sound waves almost simultaneously. Furthermore,

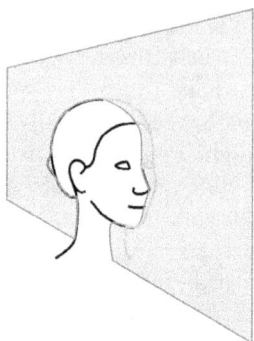

The median plane (mid-sagittal plane)   A horizontal plane at ear level

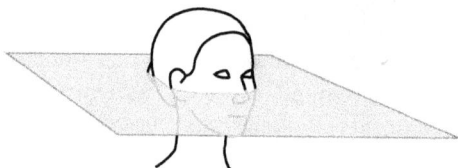

Figure 6.3 Median and horizontal planes. Sagittal planes (not shown here) are any vertical planes parallel to the median plane.

even if the sound is offset to one side in any *sagittal plane*, the ITD cannot distinguish between up and down or front and back. Therefore, ITD is useful mainly with left–right localisation. Also, when the sound frequency increases, the wavelength gets shorter and thus the phase difference between ears gets more difficult to perceive. Hence, ITD can provide clues of the sound position mainly for frequencies under 1500 Hz (Akeroyd, 2006).

In addition to ITD, there is a sound pressure difference between the two ears, the ILD. When a sound is produced on the left of the user, the left ear receives the sound waves without obstruction, but due to the *acoustic shadow* created by the head, the sound intensity gets reduced at the further ear. Whereas ITD is perceivable with the lower frequency spectrum of sounds, the opposite is true with ILD that works better with higher sounds; low frequency sounds with a longer wavelength diffract through obstacles with the size of the head, but with shorter wavelengths the head starts to have an occluding effect. Small deviations are also produced by torso, shoulders, and ear pinnae with its cavities and other shapes, all unique to each individual (Akeroyd, 2006).

To perceive the elevation of the sound as well as whether it is coming from behind rather than in front, the spectral colouration caused by the pinnae become important: depending on the direction, the sound gets scattered and reflected, leading to slightly different resonances (Akeroyd, 2006; Xie, 2013). Also, studies suggest that small head movements help to distinguish whether the sound is coming from front or back, up or down (Xie, 2013). Hence, implementing head-tracked reproduction helps with vertical localisation. Still, the resolution of spatial hearing along this sagittal plane is less accurate than on the horizontal plane where the described interaural differences are available (Blauert, 1997).

As discussed in Chapter 2, the way in which a sound is influenced by the head, torso and pinnae is described in a head-related transfer function (HRTF). HRTFs are highly individual, which is why knowledge of a user's individual HRTFs is required for perfect elevation perception. However, some features are common to many people's HRTFs. This leads to the observation that the sense of elevation can be manipulated by sound design: boosting higher frequencies, for instance high-mids (1–5 kHz), may enhance the perception that the sound is elevated (Lee, 2022; Kadel, 2024). However, cognitive aspects are similarly important, utilising listener's expectations: thunder is associated above (Larsson et al., 2010) and the metro below, even if they have similar frequency attributes. Further, it seems that semantically meaningful and easily identifiable sounds are generally easier to localise, and their locations are more likely to be remembered later (Taevs et al., 2010).

Interestingly, while the human hearing is rather good in detecting directions, it is very poor in estimating distances (Sunder, 2022). Distance hearing is based on multiple cues, such as loudness of the sound when the sound source is familiar, high-frequency absorption by air for long-distance sounds, ILD for near-field sounds, and—very importantly— reverberation (Xie, 2013). Visual cues are obviously very important, too (Brandenburg et al., 2023).

6.3 Audio paradigms

Before diving into the creation of virtual sounds, it may be helpful to revisit the channel-based, object-based, and scene-based audio paradigms discussed in Chapter 2. In many AAR production pipelines and virtual audio processes, these approaches are not used in isolation but rather combined, often working simultaneously to complement each other at different stages of the workflow.

6.3.1 Channel-based audio

Channel-based audio is the most traditional method for audio production and delivery. In this method, the final output has a fixed number of channels (stereo, 5.1, 7.1, etc.), and audio elements are mixed and panned to fit these channels. Channel-based audio is the primary format in radio, cinema, and music production, where sound is traditionally mixed for specific loudspeaker configurations.

Static binaural audio is also channel-based, as it uses two distinct channels dedicated to each ear in recording, processing, and playback. In AAR, channel-based approaches are very useful for various production phases. In runtime, they can be used for head-locked sounds including narration, music, static ambiences, and first-person sounds.

6.3.2 Object-based audio

Instead of assigning sounds to fixed channels, in object-based audio (OBA), audio elements are treated as individual objects in a 3D space with metadata that defines their position, movement, and other spatial attributes. In playback, the OBA rendering software uses the metadata to place the object audio signals across the available loudspeakers or a binaural output.

OBA is widely used in 3D video games, VR, and interactive installations, and it is also the main approach for most AAR applications. That is because virtual audio sources can be dynamically moved and easily manipulated in runtime by manipulating the metadata. Popular game engines, including Unreal Engine and Unity, often have a built-in OBA renderer. OBA is also the basis of many modern surround sound technologies including *Dolby Atmos, DTS:X,* and *MPEG-H.*

6.3.3 Scene-based audio (Ambisonics)

Scene-based audio, namely Ambisonics, takes an entirely different approach by capturing and encoding the full 360-degree sound field as a spherical representation, independent of any specific playback system, and without an explicit description of the objects that may be contained in it. In recording, a specialised Ambisonic microphone with multiple microphone capsules can be used to capture audio from all directions. The Ambisonic field can also be created in software from individual sound sources. Once such a scene has been rendered, it is not easy to manipulate or move its individual sound elements. Therefore, Ambisonics is best suited for 'baked' background ambiences that do not change or as an intermediate format, as described later. However, the Ambisonic scene can be rotated to respond to head tracking.

The bare-bones form, *first-order Ambisonics* (FOA), uses four audio channels and offers a limited spatial resolution that is typically not sufficient for reproducing a plausible auditory scene. *Higher-order Ambisonic* (HOA) formats utilise more directional components thus enabling a more accurate spatial image as well as larger sweet spot. Consequently, the amount of audio channels increases. For instance, the rather popular third-order Ambisonics uses 16 channels.

High-grade Ambisonic microphones are expensive, and with HOA formats, the number of audio channels increases, which may be an issue in some delivery pipelines. An alternative is to use parametric methods where the main audio signal is recorded using only a single good-quality monophonic microphone—which are affordable and easily available—while the spatial parameters are extracted from any lower-quality Ambisonic microphone (Gonzalez, McCormack and Politis, 2024).

If the real-world scene is captured with multiple HOA microphones, these recordings can be combined to create a volumetric sound scene where the user can traverse with 6DoF (McCormack et al., 2022; Politis et al., 2023). Such a multi-point HOA technique is supported by, for example, Zylia with their own microphones as well as the recently launched *MPEG-I* immersive audio standard.

6.3.4 Combinations

A typical virtual audio project combines these three approaches. An endless number of permutations is possible, with one example being as follows. To create an immersive soundscape for a scene, the real-world ambience is recorded with an Ambisonic microphone, in our example a second order microphone with 9 microphone capsules. Individual sounds in the soundscape are captured with normal mono and stereo microphones—or acquired from sound effects libraries. For additional depth, a pre-recorded 5.1 ambience track is also obtained.

In a *digital audio workstation* (DAW), these recordings are combined as linear audio tracks and panned to a 5.1 layout using spatialisers and encoder plugins. In our example, this 5.1 mix is finally converted to HOA, in this case third-order Ambisonics, and exported as a 16-channel audio file (see Figure 6.4).

In an authoring tool, such as a game engine or audio middleware, this HOA ambience track is combined with object-based dialogue and spot sounds that follow real-world persons and objects in a three-dimensional space. Narration and music get played as head-locked mono and stereo tracks.

In runtime, all the aforementioned sounds are processed through their own signal chains and fed to the listener's headphones. In our example, the OBA sounds, dialogue and spot sounds, get spatialised and auralised to match with the real environment. This auralised sound field is then converted into HOA for head-based rotation before getting decoded into a binaural signal (see Figure 6.5).

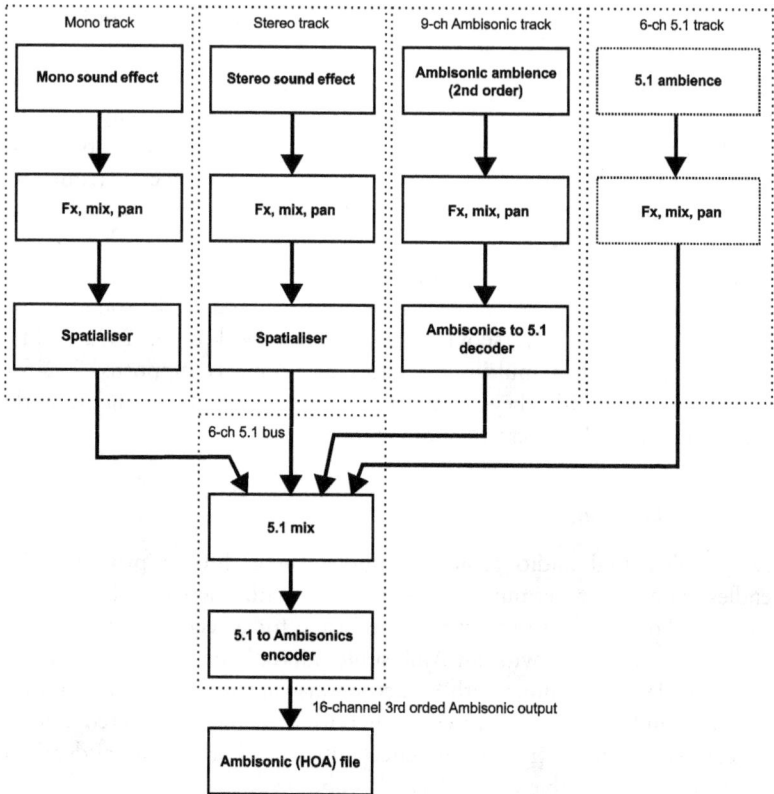

Figure 6.4 A simplified example of combining different audio types to create an ambience track in higher-order Ambisonics (HOA)

6.4 Virtual acoustics engine (spatialiser)

When using the OBA approach, the individual sounds need to be *spatialised* so that they appear as emanating from specific points in the environment. The main element of spatialisation is *auralisation*, which emulates how the given space would affect the sound waves when they propagate to the listener, and how the listener's head orientation, ears and other anatomic features would change the sound before it enters the ears. The closer this auralisation manages to create a match with real sounds, the better the chance is that the virtual sound appears as coexisting in the environment. In many AAR systems, auralisation takes place inside a virtual acoustics engine, often called the spatialiser.

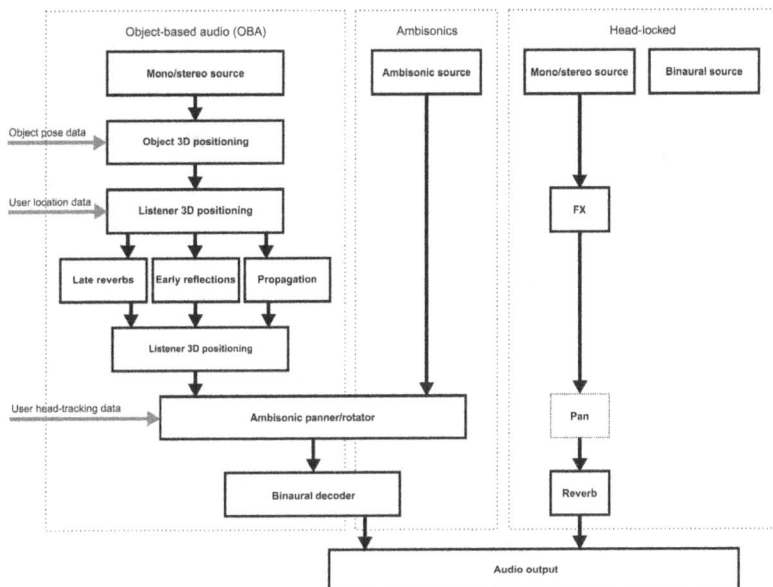

Figure 6.5 A simplified example of signal flows with different audio types in a binaural 6DoF AAR system. Especially for OBA, this is just one of many available options.

6.4.1 Sound source

The process begins with the virtual sound source that can be a pre-recorded audio file, a live audio stream, or a synthesised sound. The sound needs to be 'dry', that is, without traces of room acoustics. Otherwise the sound will appear ghostly and blurry when it gets spatialised within the environment. Using an authoring tool such as a game engine, the virtual sound is given position coordinates that align with a position in the real-world environment. For demonstration, let us attach a virtual 'meow' sound on a real cat standing in the corner of a room.

The cat object in the game engine is assigned with a monophonic audio clip. If the real cat is about to move at some point, we must track its movements and update the virtual object's coordinates accordingly. If the cat walks through a door and gets obstructed by the wall, the software can take that into account by cutting out high frequencies and applying other filters. Some systems may create a phantom sound source at the door opening to emulate the diffraction of sound waves around

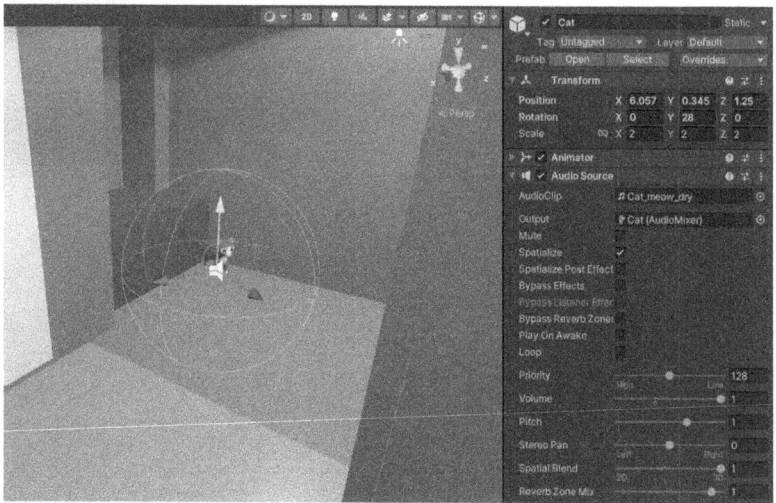

Figure 6.6 An 'avatar' of the cat in a game engine with assigned sound clip and coordinates

the door frame. Alternatively, the auralisation process can take care of that natively, depending on how the virtual acoustic system is designed.

If the user moves, the virtual environment inside the game engine—including the cat and the geometrical features of the room—need to be moved correspondingly so that their positions keep aligned with the reality.

6.4.2 Room acoustics matching

For simplicity, let the cat and the user remain stationary. Next, the sound gets auralised. In this process, the virtual sound is made audible as if the sound waves were propagating through real space, getting reflected and absorbed by walls and other surfaces. The acoustic characteristics of a space can be captured in *room impulse responses* (RIRs) which are physical-mathematical models of the sound field at any measured point. They are, in a way, acoustic 'fingerprints'. A collection of several RIRs taken from various locations are called *spatial room impulse response* (SRIR), providing three-dimensional information of the space (Meyer-Kahlen, 2024).

By convolving the original, dry sound with the SRIR, the sound gets 'treated' with the room's acoustics and appears as actually being there. Room acoustic parameters with reverb time and direct to reverberant energy ratio can be extracted from the RIRs to control a synthetic reverb effect.

There are three main ways to obtain the room's SRIR: simulating, measuring, and estimating.

6.4.3 Simulation

In simulation, the room's geometry and surface materials are first measured and defined as accurately as possible. After that, the simulation is typically realised either using *wave-based modelling* or through *geometrical acoustics* (GA) (Savioja and Svensson, 2015). Wave-based modelling simulates the propagation of sound waves across the space. While this provides accurate results, it is computationally expensive and thus likely not suitable for real-time rendering in AAR applications. It is, however, possible to pre-render the wave fields, similar to pre-baking lighting in video game production. For example, *Microsoft Project Acoustics* is a game development plugin providing such an approach. The recent PLATONE project experimented with the use of *Project Acoustics* as a semi real-time solution for AAR in urban settings (Orsholits et al., 2024).

In the GA-based approach, or 'acoustic raytracing', sound is treated as rays that reflect from surfaces and finally reach the listener. While this works with high frequencies, errors increase with low frequencies as the rays cannot simulate wave phenomena (Savioja and Svensson, 2015). Still, as this approach is computationally relatively light, it is used in many audio game engine plugins such as *Meta XR Audio SDK*.

6.4.4 Measuring RIRs

Simulation is the only way to auralise places that do not exist, like buildings that have not yet been constructed, or virtual environments in video games and VR. AAR, on the other hand, happens in real places, so for some applications, the SRIRs could be acquired by actually measuring the actual space (Meyer-Kahlen, 2024). To do this, a measurement signal, usually a sine sweep, is played back from a loudspeaker, and one or multiple microphones some distance away are used to record it. When the microphone signals are *deconvolved* with the measurement signal, RIRs are procured. The loudspeaker position, in a way, represents the sound source, and the microphone the listener. Hence, one measurement would only allow auralisation between these two points.

6.4.5 Hybrid solutions

In 6DoF applications, the listener has freedom to move anywhere within the space and there may also be multiple sound sources that can also change places. Therefore, a dense grid of RIR measurements with various speaker and microphone combinations would be required. This quickly escalates to a vast number of RIR measurements, which would be impossible to arrange in most cases, especially for mobile AAR applications; further, storing, loading and processing the huge amount of data in real-time would be too much for most systems (Meyer-Kahlen, 2024). Hence, hybrid systems have been developed that combine different approaches to optimise results and performance. For example, Brandenburg Labs' proprietary *Deep Dive Audio* technology combines data from RIR measurements with GA-based simulation, yielding highly plausible results. The *Steam Audio* plugin for game engines, in turn, combines GA with wave-based simulation. Using a more robust approach, the *dearVR* spatialiser, available as a game engine plugin, combines simple 'shoebox' reverbs with geometry-based simulation for mimicking early reflections.

6.4.6 Estimation

For mobile AAR, it is obviously not possible to set up even one measurement microphone-loudspeaker pair and run loud sweeps. Therefore, efforts have been made to find ways to predict the RIRs of a given place on-the-fly. Some approaches use computer vision and LiDAR sensors to measure room geometry and guess the acoustic properties of each surface. This technique became widely recognised through devices such as the *Apple Vision Pro* MR headset. Although it is not possible to *know* the acoustic properties of each material purely from vision, this technique may be good enough for many applications. Another approach is to analyse the reverberation caused by existing sounds in the room, called *blind estimation*. While this seems a promising solution for better plausibility compared to the visual analysis, the research is still ongoing in this front and there are no ready solutions for AAR creators at the time of writing this.

6.4.7 Binaural decoding

Now that the cat's virtual meow has been convolved by the SRIR, the sound contains the correct room reflections and colourisation, mimicking the sound wave propagation from the corner of the room to the listener. Next, one needs to simulate how the sound enters the user's ears. The head orientation obviously must be taken into account, tracked by sensors, together with the effect of head, pinnae, and torso. As discussed in Chapter 2, this anatomic information is stored in the head-related transfer functions (HRTFs).

One way to proceed is to encode the SRIR into Ambisonics. This Ambisonic field can then be rotated according to the user's head orientation. This rotated field is then translated to binaural stereo using the HRTFs. In game engines, the binaural decoder is often integrated in the virtual audio plugin, but it can also be a separate component.

6.4.8 Externalisation

The quality of binaural decoding is frequently measured in terms of its capability to *externalise* the sound. Externalisation refers to creating an auditory event that appears to emanate from outside the listener's head. In headphone listening, sounds often tend to appear as originating from inside the listener's head, called *inside-the-head locatedness* (IHL). There are three main factors that may improve externalisation: the presence of spatial cues and sense of space (e.g., reverberation), individualised HRTFs, and movement within the space (Roginska, 2018). In most AAR systems, measuring each user's personal HRTFs is nearly impossible especially in museums and other applications with rapid visitor turnover. Therefore, generic HRTFs are often used, potentially degrading externalisation and spatial accuracy for some listeners. Luckily, however, it seems that individualised HRTFs are not necessarily needed for plausible externalisation provided that the other cues are working well, including users' 6DoF movement and congruence between virtual and real room acoustics (Brandenburg et al., 2023). Also, individualised HRTFs are not that critical in reverberant spaces compared to anechoic environments (Best et al., 2020).

6.5 Conclusion

There are numerous elements required to make virtual audio seamlessly blend with reality. Moreover, each spatialiser software tool seems to rely on its own unique approach and combination of technologies. For individual AAR creators, navigating this jungle of options and workflows can feel overwhelming, and testing different systems may be time-consuming. Until widely accepted standards emerge that deliver both high-quality results and optimised performance, creators may have no choice but to continue exploring and learning. Fortunately, perfect virtual audio rendering is often unnecessary, as other factors like rich interactivity or narrative immersion (see Chapter 3) can compensate for technical imperfections.

That said, when virtual sound becomes indistinguishable from reality, the effect can be truly magical.

References

Akeroyd, M.A. (2006) 'The Psychoacoustics of Binaural Hearing', *International Journal of Audiology*, 45 (supl.1), pp. 25–33. Available at: https://doi.org/10.1080/14992020600782626.

Best, V.et al. (2020) 'Sound Externalization: A Review of Recent Research', *Trends in Hearing*, 24, p. 2331216520948390. Available at: https://doi.org/10.1177/2331216520948390.

Blauert, J. (1997) *Spatial Hearing : The Psychophysics of Human Sound Localization*. Rev. edn. MIT Press.

Brandenburg, K. et al. (2023) 'Implementation of and Application Scenarios for Plausible Immersive Audio Via Headphones', in *Audio Engineering Society Convention 155*, Audio Engineering Society. Available at: https://www.aes.org/e-lib/browse.cfm?elib=22308 (Accessed: 25 November 2023).

Gonzalez, R., McCormack, L. and Politis, A. (2024) 'S 3 MASH: Spatial Sound Scene Matching using Single-Channel Audio', in 2024 AES 5th International Conference on Audio for Virtual and Augmented Reality, Audio Engineering Society. Available at: https://www.researchgate.net/publication/384051910_S3MASH_Spatial_Sound_Scene_Matching_using_Single-Channel_Audio (Accessed: 27 November 2024).

Hudspeth, A.J. (1997) 'How Hearing Happens', *Neuron*, 19 (5), pp. 947–950. Available at: https://doi.org/10.1016/S0896-6273(00)80385-2.

Kadel, O. (2024) 'The Evolution of Spatial Audio in Immersive Storytelling', *MPSE Wavelength*, June, pp. 72–83.

Kim, S. (2015) 'Bio-Inspired Engineered Sonar Systems Based on the Understanding of Bat Echolocation', in *Biomimetic Technologies: Principles and Applications*. Cambridge, United Kingdom: Elsevier Science & Technology. Available at: http://ebookcentral.proquest.com/lib/aalto-ebooks/detail.action?docID=2102159 (Accessed: 10 April 2021).

Larsson, P. et al. (2010) 'Auditory-Induced Presence in Mixed Reality Environments and Related Technology', in E. Dubois, P. Gray, and L. Nigay (eds) *The Engineering of Mixed Reality Systems*. London: Springer (Human-Computer Interaction Series), pp. 143–163. Available at: https://doi.org/10.1007/978-1-84882-733-2_8.

Leaver, A.M. (2024) 'Perceptual and Cognitive Effects of Focal tDCS of Auditory Cortex in Tinnitus', *medRxiv*, p. 2024.01.31.24302093. Available at: https://doi.org/10.1101/2024.01.31.24302093.

Lee, H. (2022) 'Psychoacoustics of Height Perception', in J. Paterson and H. Lee (eds) *3D Audio*. New York, NY: Routledge (Perspectives on music production), pp. 82–98.

McCormack, L. et al. (2022) 'Object-Based Six-Degrees-of-Freedom Rendering of Sound Scenes Captured with Multiple Ambisonic Receivers', *Journal of the Audio Engineering Society*, 70 (5), pp. 355–372. Available at: https://doi.org/10.17743/jaes.2022.0010.

Meyer-Kahlen, N. (2024) *Transfer-Plausible Acoustics for Augmented Reality*. PhD thesis. Aalto University. Available at: https://urn.fi/URN:ISBN:978-952-64-1913-8.

Orsholits, A. et al. (2024) 'PLATONE: An Immersive Geospatial Audio Spatialization Platform', in The 2nd Annual IEEE International Conference on Metaverse Computing, Networking, and Applications (IEEE MetaCom), Hong Kong.

Politis, A. et al. (2023) 'Wide-Area 6DOF Rendering of Multi-Point Ambisonic Recordings Based on Interpolation of Spatial Parameters', in *2023 IEEE Workshop on Applications of Signal Processing to Audio and Acoustics (WASPAA)*. New Paltz, NY, USA: IEEE, pp. 1–5. Available at: https://doi.org/10.1109/WASPAA58266.2023.10248142.

Roginska, A. (2018) 'Binaural Audio Through Headphones', in A. Roginska and P. Geluso (eds) *Immersive Sound: The Art and Science of Binaural and Multi-Channel Audio*. Routledge.

Savioja, L. and Svensson, U.P. (2015) 'Overview of Geometrical Room Acoustic Modeling Techniques', *The Journal of the Acoustical Society of America*, 138 (2), pp. 708–730. Available at: https://doi.org/10.1121/1.4926438.

Sunder, K. (2022) 'Binaural Audio Engineering', in J. Paterson and H. Lee (eds) *3D Audio*. New York, NY: Routledge (Perspectives on music production), pp. 130–159.

Taevs, M. et al. (2010) 'Semantic Elaboration in Auditory and Visual Spatial Memory', *Frontiers in Psychology*, 1. Available at: https://doi.org/10.3389/fpsyg.2010.00228.

Veronese, S. et al. (2023) 'Ten-Year Follow-Up of Auditory Brainstem Implants: From Intra-Operative Electrical Auditory Brainstem Responses to Perceptual Results', *PLOS ONE*, 18 (3), p. e0282261. Available at: https://doi.org/10.1371/journal.pone.0282261.

Wenzel, E.M., Begault, D.R. and Godfroy-Cooper, M. (2018) 'Perception of Spatial Sound', in A. Roginska and P. Geluso (eds) *Immersive Sound: The Art and Science of Binaural and Multi-Channel Audio*. Routledge, pp. 5–39.

Xie, B. (2013) *Head-Related Transfer Function and Virtual Auditory Display*, 2nd edn. J. Ross Publishing.

7   Technical components

As we have learned, there are multiple ways to augment the real world with virtual audio. One of the most straightforward AAR systems would entail a hidden loudspeaker and an audio player, possibly triggered by a proximity sensor, like onboard *Barque Sigyn*. Immersive AAR experiences can also be realised with simple headphones-based systems. For instance, the *Alcatraz Cellhouse Audio Tour* uses an MP3 player and a pair of headphones, without head or location tracking, relying on the user to follow the audio guidance in sync with the narrative.

However, for more dynamic and interactive binaural experiences, the system requirements quickly increase. This chapter describes the key components of a typical interactive and wearable AAR application utilising 6DoF, such as one employing headphones, smart glasses, or other wearable auditory displays. Yet, many of the components are applicable to other platforms including direct augmentation, CTC, and WTF, as well as wearable experiences with fewer degrees of freedom. Pose tracking methods and headphones will be receiving more attention than other components. That is because computers, software, and authoring tools suitable for AAR share similar requirements with video games and VR, and are already widely discussed in other sources. Virtual audio engine, in turn, was already discussed in Chapter 6.

7.1 Wearable AAR system

A typical wearable AAR system can be understood by categorising its basic components as 1) sensors and inputs, 2) computer, 3) hardware audio interface, and 4) audio display and other outputs.

DOI: 10.4324/9781003627289-7

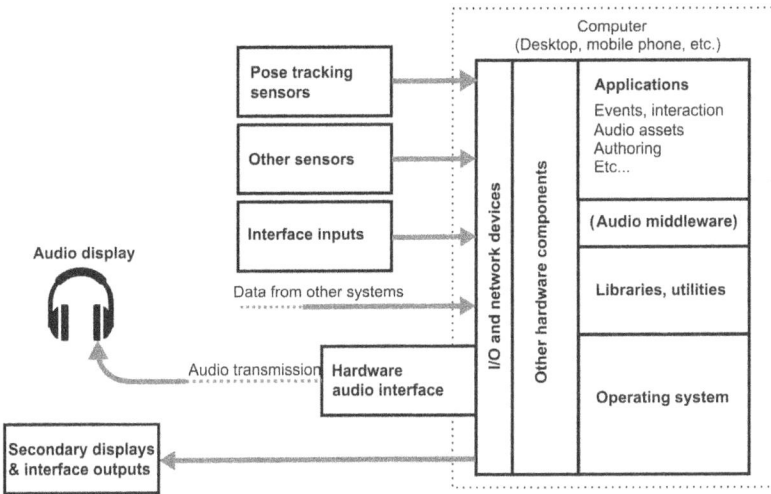

Figure 7.1 A simplified overview of the components of a wearable AAR system

7.2 Sensors and inputs

Most AAR systems are dependent on sensors that detect users' pose, movements, voices, and intentions. Other people, objects and environmental conditions can also be tracked. The sensory data is used to control the interactive logic within the system, be it shifting of perspective when the user turns their head, steering a game-like flow of events, or feeding behavioural patterns of AI-based virtual characters. Data, including captured sounds, can also be processed for various needs: recognising voice commands, enhancing speech, isolating sounds, manipulating them, making estimations of the room acoustics, to mention only a few.

The system can also include secondary interfaces with inputs such as volume buttons or touchscreens—or more creative interfaces like the wooden cube used in *ec(h)o* (Hatala and Wakkary, 2005) or mobile device held as a shield in *Audio Legends* (Rovithis et al., 2019). Signals and data can also be received from other systems in the environment, for example, to sync audio in headphones to any nearby video projection, like in *David Bowie Is* or *Dimensions of Sound*.

7.2.1 Pose tracking

Tracking the *pose* of an object means tracking its location and orientation in physical space. In headphone-based applications with 3DoF (head tracking

Table 7.1 A list of typical features that an AAR system can detect (see Lombard and Ditton, 1997; Naphtali and Rodkin, 2019; Arena et al., 2022; Yang, Barde and Billinghurst, 2022)

Source	Features
User	Pose (location and orientation) of head and other body parts
	Gestures
	User-generated sounds (voice, footsteps, etc.)
	Biodata
Environment	Pose of other people, creatures, objects
	Sounds of other people, creatures, objects
	Room acoustics
	Environmental conditions (weather, temperature, etc)
	Signals and data from other systems (video timecode, etc.)

only) or 6DoF (head and location tracking), this is crucial in order to keep the virtual sounds fixed to their three-dimensional positions regardless of users' movements. In addition to updating the auditory perspective, pose tracking is often used as the primary input mechanism for interaction, especially if no buttons, voice recognition, or other input methods are used (Yang, Barde and Billinghurst, 2022). For example, the user's absolute location in the space, or alternatively the relative proximity to objects, can trigger interactive transitions.

Head orientation is sometimes used as an interactive trigger, e.g. reacting to user's nods and shakes (e.g., Gampe, 2009)—a feature found in some smart headphones and glasses—or detecting user's head direction for assuming the direction of gaze. An interesting use for head orientation would be in a situation when it is needed to know when another person is *not* looking at the user. In *The Reign Union*, this is occasionally used to augment the other person with small non-verbal vocal sounds.

Pose tracking can also bring addition interactional benefits to any application even if the auditory display does not require it. Tracking of user's gestures, such as pinching with fingers, is becoming an important input method for AR and spatial computing devices when accurate interaction is needed.

In addition to the user's head and other body parts, other objects, people, and creatures can be tracked, too. That would enable, for instance, attaching virtual sounds on moving and movable objects. The opening angle of a door can be measured to control an audio low-pass filter, thus creating a plausible effect of sound first being obstructed and then leaking through the gap when the door opens.

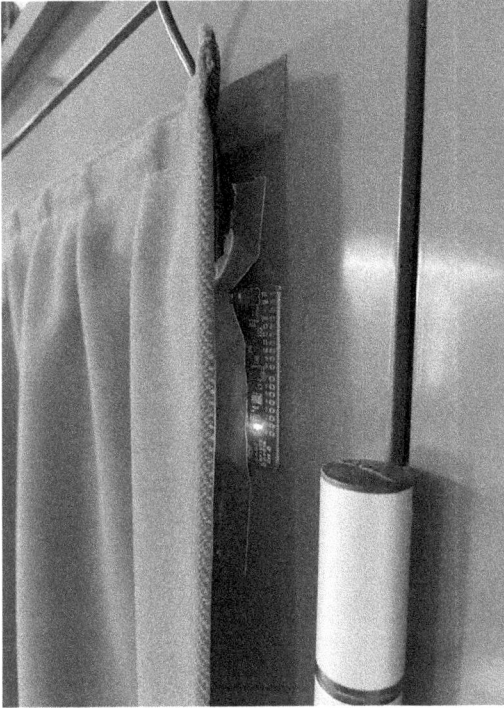

Figure 7.2 A concealed IMU sensor (*Pozyx Developer Tag*) measuring the door
   opening angle
Source: Photograph by the author

There are two main functional approaches to tracking: the system
can deliver absolute data, or relative data with changes from the last
state (Mazuryk and Gervautz, 1999). Most of the tracking systems
applicable to AAR deliver absolute data, such as systems based on fixed
anchors or cameras, or using satellite navigation. However, for
instance, a gyroscope and accelerometer provide relative data, and
hence require calibration more often due to drifting. To enable real-
time calibration when needed, and improve overall accuracy, reliability,
and robustness, different sensor types are often combined. This is
referred to as *sensor fusion*. Sensor fusion can be integrated in the
measuring unit or a device, or it can be realised by the user by combining the data from several tracking systems.

Since there are multiple technologies available for pose tracking, a
brief presentation of some of them follows with evaluation on their
usability for AAR.

7.2.2 *Optical tracking*

Optical tracking is an absolute method that uses video cameras to determine the location, and sometimes orientation, of an object (Wu, Tang and Lee, 2018). There are two primary approaches to optical tracking, *outside-in* and *inside-out*. The outside-in principle uses fixed cameras installed in the environment that track the moving target, whereas in inside-out tracking, the cameras are attached on the moving object itself, and based on what they see in the surroundings, the tracking unit tries to determine its location and orientation (Bimber and Raskar, 2005; Vertucci et al., 2023).

Optical tracking can utilise traditional 2D cameras while depth and stereo cameras offer additional accuracy. Optical tracking, especially marker-based, can be very accurate (even 0.1 mm), although it requires *line-of-sight* (LOS) and optimal lighting conditions (Ungi, Lasso and Fichtinger, 2015). Therefore, it may not be feasible for multi-user experiences where people may be blocking each other or venues that have natural light.

The advantages of optical tracking are flexibility and ability to track both position and orientation. Also, using algorithms such as *SLAM* (simultaneous localization and mapping), *PTAM* (parallel tracking and mapping), and *DSLAM* (dynamic SLAM), it is possible to determine the object positions even in unknown places (Wu, Tang and Lee, 2018; Vertucci et al., 2023). This opens up huge possibilities for content creation: experiences can be made transferable and adaptive to any environment without the need to tie them to a fixed layout or map.

Marker-based

In marker-based tracking, the camera looks for visual patterns attached to either trackable objects (outside-in) or placed in the environment (inside-out). Markers can be a constellation of reflectors illuminated with infrared (IR) light like in the *OptiTrack* system, or fiducial markers visible in normal light such as *ArUco* and *QR* codes with patterns of black-and-white shapes. By comparing them with the original marker models in the computer's memory, the algorithm calculates the object's distance, position and orientation (Vertucci et al., 2023). Even though the 6DoF information can be derived using only one camera, more cameras from different angles increase the accuracy significantly.

Markerless

As the name suggests, the markerless method does not rely on markers, but instead, the algorithm compares visual features between the real object or environment and the dataset it has been taught with. The method usually benefits from machine learning. The markerless approach allows tracking in new and unknown environments, and tracking of any object that the system is able to recognise, without the need to equip the object with markers (see Figure 7.3). Markerless tracking requires more computing power than marker-based tracking, typically causing reduced accuracy, longer latencies and lower sampling frequency. Some systems utilise graphical processing units (GPU) with parallel computing capabilities to tackle the AI-driven tracking tasks. Markerless tracking may also encounter issues when dealing with materials that have uniform, reflective, or transparent properties (Vertucci et al., 2023).

Optical, markerless inside-out tracking is used in multiple devices from smart phones and MR headsets to robotic vacuum cleaners and autonomous cars. The problem with using a mobile phone as tracking device is the device's size and weight—an inconvenience when trying to mount them on headphones—and high price. MR headsets, on the other hand, are not optimal for AAR due to their bulky form factor and obstructing display in front of eyes. However, dedicated, sole-purpose trackers are emerging, *VIVE Ultimate Tracker* being one example, released in 2023. While targeted

Figure 7.3 Markerless tracking of a picture frame using object detection with the *ZED SDK*
Source: Photograph by the author

for VR body tracking, the device is capable of autonomous pose tracking in 6DoF using two cameras and IMU. Even though its small tracking area and limited API hinders its use for AAR, it is nevertheless promising news for independent AAR developers struggling to find a convenient indoor tracking solution.

LiDAR

LiDAR (Light Detection And Ranging) is a distance measuring technology that sends out laser pulses and measures the time until they get reflected back from surfaces. This data is used to create a high-resolution, three-dimensional map of an object or an environment. LiDAR can be tremendously useful for AAR systems in accurately measuring the geometry of the environment for both user tracking as well as room acoustics estimation. It can support or replace optical tracking which often suffers from in-optimal lighting conditions (Schmalstieg and Hollerer, 2016). LiDAR sensors can be found in some smart phones and MR headsets.

Lighthouse

VIVE's proprietary *Lighthouse* tracking system is widely used in many experimental AAR projects. It uses base stations that emit precisely timed infrared laser sweeps, which are detected by sensors on tracker units. By analysing the timing and pattern of these laser flashes, the system calculates the object's location and orientation in space (HTC, 2023). The trackers incorporate IMUs which provide accelerometer and gyroscope data to refine tracking accuracy, especially during fast movements or when the IR signal is temporarily lost.

While *Lighthouse* provides fast and accurate tracking, the trackable area is limited, and the system is sensitive to bright light, which reduces its usability. There are some solutions where multiple base stations can be used together to cover larger areas, but those systems are not very stable and are difficult to set up.

7.2.3 RF-based tracking

There are a number of different radio frequency (RF) based methods for location and orientation tracking. They use radio frequency signals to determine the position of objects or devices by measuring parameters like signal strength, time of arrival, or angle of arrival.

UWB

UWB (ultra-wide band) tracking uses fixed-position 'anchors' that send radio beacon signals in the gigahertz spectrum to the environment, and a 'tag' on the trackable item, receiving signals from the anchors and using them to calculate its location in a three-dimensional space (Dardari, Closas and Djurić, 2015). In optimal conditions, UWB can reach decimetre tracking resolution, especially if height tracking (z axis) is omitted. However, in practice, accuracy and tracking reliability may be significantly reduced if anchors are installed too close to users or if walls and other structures interfere with the radio signals. One benefit of UWB is scalability, allowing numerous tags to track themselves simultaneously, an important feature in museums and other public applications.

AoA

Bluetooth Angle of Arrival (AoA) is a location tracking method that determines the direction of a Bluetooth signal by analysing the phase differences between multiple antennas in an array. By combining this directional information with signal strength or triangulation from multiple receivers, the system can estimate the position of a device. Basically, any Bluetooth device can be tracked, and dedicated tags are very light and small. However, multiple antennas installed on the ceiling are usually required, which may not always be possible in, for example, historical buildings. Also, low installation height reduces accuracy significantly, and like all RF-based systems, AoA is sensitive to signal reflections in cluttered environments causing tracking errors.

Other RF-based tracking technologies

Several other short-range wireless radio communication standards have also been used for location tracking purposes, even though they have not been designed for that. Common ones are Wi-Fi, RFID, NFC, and ZigBee. Their accuracy is usually considered poor, about 1–5 metres, and they present other challenges, too, including reduced possibilities for scalability (Dardari, Closas and Djurić, 2015). However, there is some research on using WiFi signals and deep learning architectures for accurate human pose tracking through machine learning, avoiding the occlusion and lighting limitations of optical systems (Geng, Huang and Torre, 2022).

The 5G technology can also be used for positioning, and in indoors settings even centimetre-level accuracy can be achieved; however, that would require large antenna arrays (Italiano et al., 2024), which so far makes this method unpractical.

7.2.4 Electromagnetic tracking

Electromagnetic (EM) tracking, sometimes called EMF (electromagnetic field) tracking, uses localised electromagnetic fields to determine the precise position and orientation of objects in 3D space, offering high accuracy and relatively low latency. However, its reliance on low-frequency fields limits its range to a few metres. Some systems enable the use of multiple transmitters to extend the coverage, but the system would still be susceptible to interference from metal structures or electronic devices. While EM tracking has been used since the 1960s especially in military and medical environments, the technology is still being developed and improved (e.g., Amfitech, 2024).

7.2.5 Satellite positioning

Satellite positioning is based on constellations of satellites orbiting the Earth. They send positioning and timing data which is received by the trackable devices. The devices use the data to interpret their location relative to Earth. Four *GNSS* (global navigation satellite systems) with their own satellites are operational, *GPS* (USA), *Galileo* (Europe), *GLONASS* (Russia), and *BeiDou* (China). Modern receivers can interpret data from all or most of them simultaneously. GNSS is made for outdoor tracking, and whereas the accuracy is normally around 1 metre at its best (Fernandez-Hernandez et al., 2018), with extensions such as RTK (real-time kinematic positioning) even sub-centimetre accuracy can be achieved (Boquet, Vilajosana and Martinez, 2024). This opens highly interesting possibilities for outdoor 6DoF AAR experiences on a global scale (see Orsholits et al., 2024).

7.2.6 Ultrasound tracking

Ultrasound tracking determines the position of an object by triangulating the time it takes for ultrasonic (ca. 19–45 kHz) sound waves to travel between a transmitter mounted on the trackable object and multiple receivers installed in the environment. Some systems support inverted arrangement where the trackable object carries a receiver and the transmitters send sound pulses on different frequencies (e.g., Marvelmind, 2023). In optimal conditions, accuracy can be on a centimetre scale, but in practice, reflections, noise, and loss of line-of-sight easily reduce the performance. Also, even though the ultrasonic frequencies are above human hearing range, some transmitters emit audible clicks that may be distracting.

Instead of using dedicated transmitters and receivers, ultrasound tracking can be attempted with regular loudspeakers and microphones (e.g., ALPS, 2018). This approach would leverage existing hardware and reduce the need for specialised equipment, potentially allowing for easier integration into diverse environments.

7.2.7 Proximity sensors

Proximity sensors are capable of detecting distance or presence, but not direction or position. Hence, they do not enable 6DoF tracking, but are discussed briefly here as they are common in many past and present exhibitions and applications.

Bederson's (1995) early AAR prototype used infrared transmitters—similar to TV remote controls—and a wearable receiver to detect when the user was close to a museum exhibit. Nowadays, IR transmitters are common for proximity detection, and additionally, user's presence can be detected by reading the heat map of the area of interest with a thermal (IR) sensor, and detecting human body heat from it.

RFID (radio frequency identification) is a common method as well, where a tag worn by the user is detected at a distance by a reader device. Bluetooth Low Energy (BLE) beacons, such as Apple's *iBeacon*, in turn, emit signals that are detected by BLE-enabled devices, like mobile phones, which then calculate their distance to the beacon. Further, proprietary systems may use any other RF mode, such as the *Sennheiser guidePORT* system at the *David Bowie Is* exhibitions that utilised transmitters operating at the 127 kHz band.

7.2.8 Inertial tracking

Inertial tracking is usually achieved using an inertial measurement unit (IMU). It is an integrated sensor module that is comprised of an accelerometer, gyroscope, and sometimes magnetometer. By combining the data from these sensors, rotation angles (pitch, roll, yaw) can be determined as well as linear acceleration. The magnetometer can be used as a digital compass. IMUs can be found in a myriad of devices ranging from smartphones to drones and head-tracking headphones to VR controllers.

An IMU with accelerometer and gyroscope is called '6DoF', and when a magnetometer is added, it is called '9DoF'. These degrees of freedom, however, are different from movement degrees. DoFs in IMU refer to how the sensor *measures* different movements, while movement tracking DoFs such as 6DoF, describes the *actual movement* of the object.

The data from a gyroscope and accelerometer is relative, so drifting tends to happen over time and thus recalibration is needed. Calibration can be attempted by fusing data from other sensors, such as cameras, that may be able to detect the actual pose of the trackable object. Yang et al. (2020) used a clever 'opportunistic recalibration' method: When their IMU drifted, the spatialised sound from an object seemed to come from a wrong direction. However, since the audio was contextually congruent with the object, the users tended to point their head at the object, not the sound. Now, they could measure the offset angle of the user' head and correct the IMU drift.

While the magnetometer module reduces drift significantly, it may behave erratically in some indoor environments due to magnetic interference from various sources. Hence, the use of 9DoF IMUs may not be possible, or the magnetometer must at least be switched off in problematic areas.

7.2.9 Other sensors

In an application where interactions with objects and environments are based on users' movements, pose tracking with possible gesture recognition may be the only required sensory method. However, additional sensors may provide data that enriches the experience greatly. Data can be gathered from the user, other users, the immediate environment, or from beyond the user's perception. Ultimately, an environment-aware system can utilise any surrounding stimuli, like brightness, temperature, or sounds (see Naphtali and Rodkin, 2019) as long as there are appropriate sensors available and means to analyse their data.

Microphones

Microphones are useful sensors for multiple purposes. They can capture the user's voice for voice commands, or transmit it to other users for communication or narrative purposes. Voice and other sounds, such as footsteps, can be recorded and fed back to the user, possibly enhanced by reverberation or other augmentations. Voice and breathing can also be analysed for creative uses or medical purposes, for instance, by monitoring a user's respiratory rate (Doheny et al., 2023).

Microphone arrays in wearable devices can be used for sound beamforming, 'zooming' into certain sounds, which aids sound and speech isolation from background noise. This opens further possibilities to attenuate and replace the sounds with virtual ones. Data for blind estimation of a room's acoustic properties—although yet in a research stage—could also be collected through a microphone array (see Meyer-Kahlen, 2024).

Unless there is access to, or the desire to use, the wearable device's integrated microphones, external microphones need to be incorporated. It must be noted, that this would require integrating an audio interface and signal transmission to the processor, with likely weight and latency issues.

Acoustic watermarking

Acoustic watermarking warrants acknowledgment in the context of AAR. It is a signal processing technique where inaudible information is embedded within an audio signal, allowing it to carry hidden data without noticeably altering the original sound (Lin and Abdulla, 2015). For instance, museum exhibits could emit watermarked sounds through loudspeakers. As visitors approach different displays, their devices could detect these embedded signals, delivering tailored audio narratives or triggering various interactional cues. This technique eliminates the need for additional hardware or visible signals. Moreover, as sound can travel around obstacles such as walls and furniture, acoustic watermarking is potentially effective in complex indoor environments.

Biodata sensors

Biodata may be useful for an AAR application as discussed in Chapter 3.3. Skin temperature, heart rate, and electrodermal activity are just some variables that can be relatively easily measured. Many activity watches, rings, and other devices feature biodata sensors, and while data access is possible via APIs or platforms, it often requires technical expertise and programming.

7.3 Computer

The computer and the software running on it form the brain of an AAR system (see Figure 7.1). Mobile phones and other computing devices are considered here as computers, which, fundamentally, they are. Simply put, the computer receives data from sensors either through physical I/O ports or network interfaces. This data is then processed by the operating system and associated software, which use it to control the progression of events within the AAR scene. The software manages audio playback, handles virtual audio rendering, and executes other functions based on rules and algorithms defined during the authoring phase. Finally, the audio interface outputs the audio signal to headphones or loudspeakers.

Many AAR applications run on either a mobile phone or a traditional computer. However, the hardware can also be a pair of stand-alone smart glasses, a 'computing platform' such as *Nvidia Jetson*, a tablet, a single-board computer such as *Raspberry Pi*, or even a microcontroller such as *Arduino*. The computer itself does not necessarily need to run any audio: in Bederson's (1995) AAR prototype, the interaction logic was programmed into a Motorola microprocessor which controlled a modified Sony MiniDisc player containing the audio recordings.

The software can be a custom-made application running on a game engine runtime environment such as *Unity* or *Unreal Engine*, or developed using programming environments and frameworks such as C++ with *OpenFrameworks*, with many additional approaches available. Middleware programmes like *Wwise* or *FMOD* are often integrated for audio authoring and rendering, while APIs and SDKs (software development kits), such as *ARCore* or *ARKit*, enable spatial tracking and interaction.

The quality of the virtual audio rendering may set high demands for the performance of the runtime engine. While cleverly designed virtual audio software can be highly efficient, in general, the more audio assets and advanced acoustic simulations are required, the more processor-intensive the application becomes. This easily becomes a problem especially on mobile devices, forcing use of higher-than-desired audio buffer sizes, which increase latency, or requiring compromises in rendering quality and the amount of virtualised audio objects.

The advantage of mobile devices, however, is their ability to function as autonomous systems, as they typically integrate sensors such as cameras, IMUs, LiDAR, microphones, and an audio output interface. In contrast, external computers can render higher quality audio and more complex auditory scenes. However, the audio must then be transmitted to the user's headphones—preferably wirelessly—introducing its own constraints, such as limiting the size of the experience area.

To reduce computational load for a single device or computer, tasks can be distributed to multiple computing units, especially in fixed, site-specific installations. For example, user tracking can be handled by several, independent units, scene interactions by one, and virtual audio by a separate computer. With mobile experiences, the wearable units can offload some tasks to a central server, such as virtual acoustics rendering (see Orsholits et al., 2024), although delays in this approach may limit the usability.

7.4 Audio interface and transmission

A hardware audio interface is a physical audio device that converts digital audio signals into analogue audio (DAC, *digital-to-analogue converter*) for playback through speakers or headphones. An audio interface can also include input devices for microphones with pre-amplification circuits and *analogue-to-digital converters* (ADCs). In many mobile devices and headphones with digital audio connections, such as Bluetooth or USB, the audio interface is integrated directly into the device.

Often, the best way to connect headphones to the computer—or a separate audio interface—would be to use a physical audio or data cable. If that is not possible or feasible, a wireless connection is needed. A Bluetooth connection to wireless headphones is an easy option, but may introduce unacceptable latency as discussed later. Professional wireless systems are also available, commonly used by musicians for in-ear monitoring (IEM). However, the size, weight, and power consumption of the receiver unit may become an issue, especially when mounted on headphones. Also, as is the case with all radio-based systems, it must be remembered that walls and other structures may cause significant losses to the signals, and even with good antennas that are properly installed, the range may remain modest.

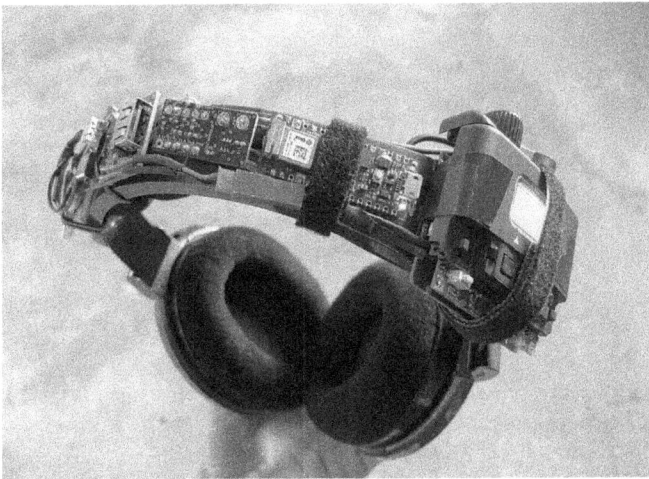

Figure 7.4 A wireless, digital audio receiver (shown on the right) mounted on AAR prototype headphones
Source: Photograph by the author

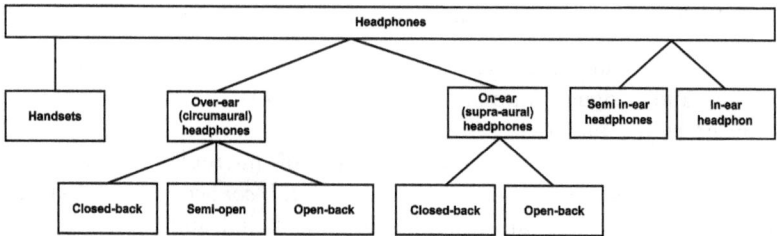

Figure 7.5 The main headphone types categorised by structure

7.5 Headphones

Various auditory displays suitable for AAR were introduced earlier in Chapter 2.10. This section takes a closer look at headphones, the primary audio output device for most AAR applications. Even though terms like 'earphones' or 'wearable listening devices' could more accurately describe these devices, for simplicity, 'headphones' is used as a catch-all term throughout this book.

Headphones offer an individualised auditory experience. Common to all headphones is the close proximity to ears enabling controllable sound quality and low sound pressure level. This consequently reduces sound leakage and disturbance to surroundings.

Headphones also come with some disadvantages. First of all, each user requires their own device, which increases expenses and maintenance demands. Additionally, some headphones remind of themselves either through their physical weight and feeling (pressing on ears, heat, etc.) or due to the imperfect sound quality. A significant challenge arises in binaural 6DoF applications, where precise user tracking is essential for accurate virtual audio spatialisation.

7.5.1 Over-ear headphones

Over-ear or *circumaural* headphones have large earcups that cover the whole outer ear. The earcups are usually attached to a headband, which is useful for mounting special devices and markers needed for the particular AAR application. There are *closed-back* and *open-back* designs. Closed-back headphones attenuate external sounds and prevent leaking from inside to out. This may be optimal in situations where environmental sounds would be too disturbing in relation to the virtual audio content. However, closed-back headphones introduce resonant or 'boomy' bass response, and as they are not ventilated, they accumulate heat when used for a long period (Tran, Amrein and Letowski, 2009).

Open-back and semi-open headphones, in turn, have a structure that 'breaths', resulting in a more open and natural sound. They are also acoustically more transparent than closed headphones, which is seen as a benefit in terms of maintaining the sense of the real surroundings and other people while improving the externalisation of virtual sounds (McGill et al., 2020). Nevertheless, in practice, closed-back design is sometimes chosen for their lower price range, and because open headphones may not, in the end, enable verbal communication with peers significantly better (Armbruster, 2024). Still, while headphones can often feel isolating, open-back designs reduce this effect and are generally considered more pleasant (Vazquez-Alvarez et al., 2016). Open-back headphones also always introduce some level of acoustic colourisation to the real-world sounds, which might be inadmissible for some critical applications (Jacuzzi, 2018). However, since hearing is adaptive and likely gets used to the coloured soundscape, it may be a good idea to try to equalise the virtual content similarly to blend these two acoustic layers better together.

7.5.2 On-ear headphones

On-ear or *supra-aural* headphones rest on the ear. They are less bulky and often lighter than circumaural headphones, but they may feel uncomfortable in the long run as they press the ears. They also come with closed and open design, although the sound isolation with closed-back on-ear headphones is not as good as with over-ear headphones.

7.5.3 In-ear headphones (earbuds)

In-ear headphones and *earbuds* nowadays mean more or less the same. In-ear headphones are placed inside the ear canal and they are sealed and secured in place with a silicon ring or even a custom-made mould. While potentially providing a very good sound quality, the problem with these devices may be discomfort after longer periods of use as well as the 'occlusion' effect that makes the own voice sound boomy and amplifies in-body sounds such as eating (Albrecht, Lokki and Savioja, 2011).

Semi-in-ear headphones, on the other hand, are placed in the concha and they do not provide any acoustic seal. They may feel a bit uncomfortable and drop off easily, and one does not see them around much anymore, apart from airplanes as disposable headphones.

In-ear headphones, and to some extent closed-back headphones, may increase user's awareness of self through hearing their own bodily sounds. This, in consequence, can contribute to feeling less connected with the surroundings (Larsson et al., 2010).

Earbuds are private devices, and for hygienic reasons, they are not usable in public applications as hand-out devices. It is possible that an AAR experience could rely on users' own devices and headphones. That, however, poses challenges for the design as there is no way to have prior knowledge about each user's end device. The level of acoustic transparency would be critical information in the context of augmented reality. The frequency response of the headphones may also be good to know, and if correct audio levels are of importance, some kind of calibration process in the beginning of the experience would be needed.

7.5.4 Open-ear headphones

Open-ear headphones are acoustically (almost) completely transparent, using drivers in close proximity to ears without an enclosing structure. Some open-ear models use a headband to support the transducers (e.g., *AKG K1000* and '*Mushrooms*') while some clip onto the ear's helix, similar to an earring (e.g., *Bose Ultra Open Earbuds*). Open-ear headphones allow users to hear real-world sounds naturally, preserving a sense of 'being part of the environment', a fundamental principle of AR and MR experiences (Larsson et al., 2010, p. 152).

Near-ear speakers use similar logic and are integrated in some smart glasses, AR glasses and MR headsets.

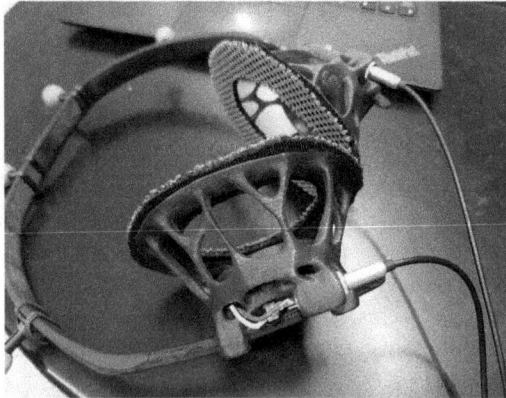

Figure 7.6 Acoustically transparent, 3D-printable headphones nicknamed '*Mushrooms*' designed by Alexander Mülleder (Mülleder et al., 2023). The headband is mounted with *OptiTrack* reflective markers.
Source: Photograph by the author

7.5.5 Bone-conduction headphones

Besides open-ear headphones and near-ear speakers, bone-conduction (BC) headphones are also acoustically completely transparent. Instead of transmitting sound waves to ears through air, they vibrate the bones of the skull which, consequently, simulate the cochlea inside the ear. Hence, there is no need to place any device in or over the ear. However, BC devices require tight pressure against the skull, potentially leading to skin problems and discomfort (Nishimura et al., 2023).

7.5.6 Cartilage conduction headphones

Besides using bone conduction, sound can be transmitted through vibrating the aural cartilage of the ear. Cartilage conduction (CC) is used in some hearing aids mainly available in Japan, and while research on them is still limited, it seems that CC can provide improved comfort and better spatial localisation compared to BC devices (Nishimura et al., 2023).

7.5.7 Hear-through

Many of the modern headphones, including earbuds, are equipped with ANC and electronic hear-through, with some models recognising speech and other sounds for adaptive soundscape manipulation. The headphone structure is usually closed, acoustically blocking out some of the surrounding sounds. They use a *pseudoacoustic* system where a microphone array is installed on the outer surface of the device, capturing the outside sounds (Härmä et al., 2004). Using digital signal processing (DSP), the environmental sounds can be attenuated by feeding inverted audio signal to ears, thus 'cancelling' the original sound.

Such a 'microphone-through' (Albrecht, Lokki and Savioja, 2011) system could be used for AAR to offer acoustic transparency, not unlike in MR headsets with a video-through AR display. Compared to an acoustically transparent AAR system, a pseudoacoustic system is more complex to build and operate, as real-world sounds must be processed and fed into the headphones. However, this approach offers the advantage of enabling the removal and replacement of the real-world soundscape, paving the way for MR experiences. Yet, if attempting to use commercial noise-cancelling headphones, the problem for an AAR developer may be that the headphone manufacturers do not usually expose their APIs in a way that the microphone signal could be 'hijacked' and manipulated.

7.5.8 Smart headphones (hearables)

When headphones are equipped with sensors and sophisticated technologies, they are sometimes called *smart headphones or hearables*. Through AI-based technologies, ANC systems can be more effective, and often adaptive, being able recognise certain sounds, such as speech, and let it through while attenuating others. Multiple microphones enable beamforming, or acoustically 'zooming' into desired sound sources, and the occlusion effect can be minimised. In addition to modern hearing aids that are capable of such, there are also consumer-level earbuds equipped with similar features, such as *Pixel Buds Pro 2*.

Some headphones, earbuds, and smart glasses also feature an integrated head-tracking sensor, like *AirPods Pro*. The wireless audio transmission used in many headphones can, however, introduce latency into the audio signal chain, making the soundscape lag behind when turning the head.

7.5.9 Smart glasses

Smart glasses refer to eyewear that is enhanced with electronics; some glasses are equipped with near-ear speakers ('audio smart glasses'), some feature see-through video displays ('AR glasses'), while some models come with cameras for taking pictures and videos (Greenwald, 2024). Many devices combine these features, and all of them usually have an array of microphones for voice commands and audio recording as well as wireless connectivity to mobile phone and WiFi. Some glasses incorporate head tracking, which would be crucial for any real AR application including AAR utilising binaural spatial audio. *Bose Frames* used to have head tracking (used in, e.g., *Please Confirm You Are Not a Robot,* see Chapter 5.7.), until the termination of Bose's AR programme in 2020 after which the glasses are now sold as regular audio glasses.

Figure 7.7 Ray-Ban Meta smart glasses with built-in microphones, near-ear
   speakers and video camera
Source: Photograph by the author

7.5.10 *Sound quality*

Higher-quality headphones can naturally reproduce virtual sounds more
realistically than lower-quality ones. However, even at the high-quality
range, no headphone model has a perfectly transparent frequency response,
thus altering the binaural reproduction. For the spatialisation to work
optimally, this headphone response should be compensated with equalisation, and preferably adjusted for each user individually as the headphone–
ear coupling may vary from person to person (Sunder, 2022). If the used
headphone model is known beforehand, as is the case in integrated systems,
equalisation can be pre-performed. However, if the system relies on users'
own headphones, proper equalisation may not be achievable. Moreover, it
appears that the ears can slightly adapt to less realistic audio quality in
headphone-based AAR systems, allowing for an acceptable experience even
with suboptimal audio quality (Engel and Picinali, 2017).

7.6 Authoring

Authoring refers to the process of programming and scripting the AAR
system; among other things, it is about defining the sensor inputs and
how that data should be used, placing the virtual sounds and other
objects in the three-dimensional space, setting their parameters, scripting
the logic how everything works together, and adjusting the virtual audio
tools to auralise the virtual soundscape as well as possible.

For some AAR environments, such as geolocated walks, there are user-friendly web and application platforms that make authoring very easy. For other interactive AAR environments, however, tools are more complex, requiring at least some level of programming. While this opens creative possibilities, it may limit fast iteration of ideas, prevent non-programmers from taking an active role, and slow down the overall development of the medium in general (Schmalstieg and Hollerer, 2016).

Game engines offer a shortcut to authoring with ready-made tools for 3D asset placement and handling, state machines, audio tools, and numerous add-ons including plugins for virtual audio processing, pose and object tracking, speech recognition, communication between devices, online connectivity, etc. Game engines free the creator from many programming tasks, enabling concentration on other areas of the creative process (Buttle, 2020). Some game engines have visual scripting interfaces (e.g., *Blueprints* in Unreal Engine and *Visual Scripting* in Unity), which not only help people less skilful in programming, but make it easier to manage complex narrative and interactive structures that would be tedious to comprehend solely through lines of code.

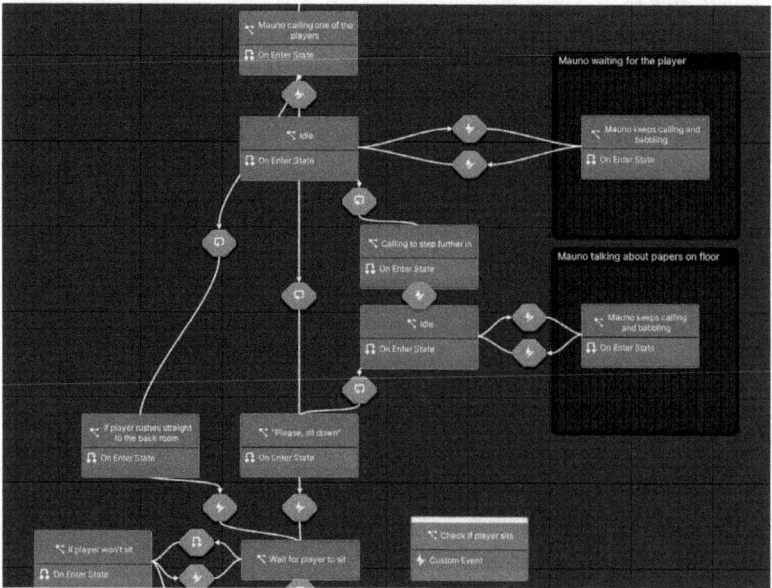

Figure 7.8 A part of the interactive story structure of *The Reign Union* in the *Unity Visual Scripting* interface

In a typical authoring process, the real-world is 'brought in' as a virtual representation, something that the audio, other content, and the user can be spatially located in. In geolocated audio walk applications a two-dimensional map is often used, while three-dimensional modelling is necessary for more complex projects. This modelling can be done by hand using measurements of the real environment, although some dynamic systems can scan the environment automatically, for instance, using cameras and LiDAR sensors. While the virtual model helps in positioning and controlling narrative elements, it can also be useful for virtual audio rendering especially if the engine utilises geometric acoustics (GA) (see Chapter 6.4).

The user can be represented as an avatar that moves within the virtual scene according to the tracking data received from sensors. The game engine's 'listener' component, which functions as virtual ears, is positioned in the avatar where the real person's ears would be. Then, it is the virtual audio engine's task to auralise the sound propagation from the sound sources to the listener—with careful manual adjustments by the user.

Interactions and other scene logics including the behaviour of the sounds are scripted using the tools available in the authoring environment. Finally, the project may be compiled into an executable application for the target platform (MacOS, Windows, Linux, Android, iOS, etc.).

7.7 Motion-to-sound latency

A head-tracked binaural system attempts at keeping the virtual soundscape spatially aligned with the real world regardless of the head movements (Roginska, 2018). However, some amount of delay is always introduced between the moment the head moves and before the user hears the virtual soundscape react to the movement. This 'motion-to-sound latency' is accumulated by tracking sensors, data transfer, audio rendering, and digital-to-analog (DAC) conversion in the audio output stage (Meyer-Kahlen et al., 2023).

The motion-to-sound latency causes problems in AAR as it makes virtual sounds and room acoustics lag behind from their real-world locations when the user's perspective changes. This off-sync can easily harm the illusion of nonmediation and sense of presence (see Lombard and Ditton, 1997)

Latencies below 60 or 50 milliseconds (ms) are usually considered as imperceptible in most virtual audio cases, while less than 30 ms being nearly unnoticeable (Brungart, Kordik and Simpson, 2006; Brandenburg et al., 2023). In practice, those numbers can be achieved if the tracking system is very fast, such as *OptiTrack* or *VIVE Tracker*, audio rendering is well optimised and running on a powerful computer with a small audio buffer, and as many connections are tethered as possible. However, many AAR applications cannot run on such specifications. For example, even if head orientation tracking using an IMU tracker is relatively fast with a 10 ms latency (Supper, 2021), the location tracking system may introduce over 100 ms latencies. Luckily, snappy head-orientation tracking is generally more critical, while fast location tracking becomes crucial only when virtual sounds are nearby.

Audio rendering adds additional delays, as discussed in Chapter 7.3. Complex auditory scenes place greater demands on the computer and software, often resulting in larger audio buffer sizes and increased latency. Finally, if wireless audio transmission to headphones is required, it may add significant latency. Professional wireless systems can achieve latencies under 1 millisecond. Some low-latency Bluetooth codecs, such as *aptX LL* and *LHDC LL*, offer delays of 30–40 ms, whereas standard Bluetooth codecs can introduce delays exceeding 200 ms, which can significantly impact the user experience (Katz, 2024). With all that, if the system can reach motion-to-sound latency of around 100 ms, even that can be considered rather good in terms of maintaining plausibility (Lindau, 2009).

The ventriloquism effect, where virtual sounds 'magnetise' to plausible physical objects, can compensate for some dynamic errors (see Ahrens, 2019). However, practical experience suggests that this tolerance is limited, and the sound cannot deviate significantly. If tracking is unstable and the virtual sound appears to drift around the physical object, the illusion is likely to break more easily.

7.8 Complete systems and standards

Although it is currently not possible to purchase a complete 6DoF AAR system anywhere, at least to the knowledge of the author, there are a few proprietary systems made with companies that produce AAR content. However, even their systems are not complete: while they pack technology in a well-thought and reliable form, the authoring pipeline or tracking system may still be relying on 3rd party tools and complex workflows.

Partial solutions are or will be available for purchase, and the first-ever standard for interactive, immersive audio, *MPEG-I*, should have been published in 2025 when this book comes out. The standard, a joint effort of multiple companies, aims to unify and stabilise audio rendering across devices. It enables 6DoF within audio scenes and supports both binaural and loudspeaker setups. MPEG-I is scalable and able to optimise rendering resolution for mobile devices with limited processing capacity. Built on MPEG-H coding, it supports channel-based and object-based audio (OBA) as well as higher order ambisonics (HOA), and introduces multi-point HOA (Herre et al., 2024). In terms of AAR, one important feature is the possibility to input parameters, such as room reverberation time (RT60), in order to match the virtual acoustics with the real environment (Herre and Disch, 2023).

Another recently published standard and codec particularly interesting in terms of AAR is *IVAS*. As a joint effort of multiple telecommunications and technology companies, the codec enables, for instance, ad-hoc capture of a spatial auditory scene—such as people around the meeting table—by using a single mobile device. Also, multiple audio streams from each participant's remote locations can be spatialised at the end user's device (Bruhn, Multrus and Varga, 2022; IVAS Codec Public Collaboration Contributors, 2023).

It remains to be seen how widely these standards will be adopted by hardware and software companies, how effectively they will simplify and unify production pipelines, and how they will enable seamless cross-platform content distribution. Still, their existence already reflects the growing recognition of the importance of virtual and spatial audio in the industry.

7.9 Conclusion

The number of possible technologies to realise AAR may seem overwhelming. Choosing the right approach is always a challenging process with a risk that testing one technology takes money and time, and in the end, turns out not to be the best. The lack of ready-made, complete systems openly available for content creators—apart from geolocated AAR platforms—is likely to slow down ideas from getting into AAR applications and experiences. This chapter has hopefully provided guidance for practitioners in selecting the right components for a given project, reducing the need for extensive trial and error, while also offering a glimpse into the technological potential and opportunities available in AAR.

References

Ahrens, A. et al. (2019) 'Sound Source Localization with Varying Amount of Visual Information in Virtual Reality', *PLOS ONE*, 14 (3), p. e0214603. Available at: https://doi.org/10.1371/journal.pone.0214603.

Albrecht, R., Lokki, T. and Savioja, L. (2011) 'A Mobile Augmented Reality Audio System with Binaural Microphones', in *Proceedings of Interacting with*

Sound Workshop: Exploring Context-Aware, Local and Social Audio Applications. New York, NY, USA: Association for Computing Machinery (IwS '11), pp. 7–11. Available at: https://doi.org/10.1145/2019335.2019337.

ALPS (2018) 'ALPS – Acoustic Localisation Positioning System'. Available at: https://alps.fi/ (Accessed: 26 November 2024).

Amfitech (2024) 'AMFITRACKTM, AmfiTrack'. Available at: https://www.am fitrack.com (Accessed: 26 November 2024).

Arena, F. et al. (2022) 'An Overview of Augmented Reality', *Computers*, 11 (2), p. 28. Available at: https://doi.org/10.3390/computers11020028.

Armbruster, S. (2024) 'Conversation with Matias Harju, March 7'.

Bederson, B.B. (1995) 'Audio Augmented Reality: A Prototype Automated Tour Guide', in *Conference Companion on Human Factors in Computing Systems – CHI '95*. Denver, Colorado, United States: ACM Press, pp. 210–211. Available at: https://doi.org/10.1145/223355.223526.

Bimber, O. and Raskar, R. (2005) *Spatial Augmented Reality: Merging Real and Virtual Worlds*. Wellesley, Massachussets: A K Peters.

Boquet, G., Vilajosana, X. and Martinez, B. (2024) 'Feasibility of Providing High-Precision GNSS Correction Data Through Non-Terrestrial Networks', *IEEE Transactions on Instrumentation and Measurement*, 73, pp. 1–15. Available at: https://doi.org/10.1109/TIM.2024.3453319.

Brandenburg, K. et al. (2023) 'Implementation of and Application Scenarios for Plausible Immersive Audio Via Headphones', in *Audio Engineering Society Convention 155*, Audio Engineering Society. Available at: https://www.aes.org/e-lib/browse.cfm?elib=22308 (Accessed: 25 November 2023).

Bruhn, S., Multrus, M. and Varga, I. (2022) 'IVAS – Taking 3GPP Voice and Audio Services to a New Immersive Level', *3GPP Highlights*, October, pp. 8–9.

Brungart, D., Kordik, A.J. and Simpson, B. (2006) 'Effects of Headtracker Latency in Virtual Audio Displays', *Journal of the Audio Engineering Society*, 54, pp. 32–44.

Buttle, P. (2020) 'The Power Behind Video Games: A Look at Game Engines', *Medium*. Available at: https://medium.com/wetheplayers/the-power-behind-vi deo-games-a-look-at-game-engines-2731315086e0 (Accessed: 3 May 2021).

Dardari, D., Closas, P. and Djurić, P.M. (2015) 'Indoor Tracking: Theory, Methods, and Technologies', *IEEE Transactions on Vehicular Technology*, 64 (4), pp. 1263–1278. Available at: https://doi.org/10.1109/TVT.2015.2403868.

Doheny, E.P. et al. (2023) 'Estimation of Respiratory Rate and Exhale Duration Using Audio Signals Recorded by Smartphone Microphones', *Biomedical Signal Processing and Control*, 80, p. 104318. Available at: https://doi.org/10.1016/j.bspc.2022.104318.

Engel, J.I. and Picinali, L. (2017) 'Long-Term User Adaptation to an Audio Augmented Reality System', in *Institute of Acoustics Proceedings. ICSV24 2017*, London. Available at: https://www.ioa.org.uk/catalogue/paper/long-term -user-adaptation-audio-augmented-reality-system.

Fernandez-Hernandez, I. et al. (2018) 'Galileo High Accuracy: A Programme and Policy Perspective', in 69th International Astronautical Congress, Bremen, Germany.

Gampe, J. (2009) 'Interactive Narration within Audio Augmented Realities', in I.A. Iurgel, P. Petta, and N. Zagalo (eds) *Interactive Storytelling*. Berlin, Heidelberg: Springer Berlin Heidelberg (Lecture Notes in Computer Science), pp. 298–303. Available at: https://doi.org/10.1007/978-3-642-10643-9_34.

Geng, J., Huang, D. and Torre, F.D. la (2022) 'DensePose From WiFi', *arXiv*. Available at: https://doi.org/10.48550/arXiv.2301.00250.

Greenwald, W. (2024) 'The Best Smart Glasses for 2024', *PCMag UK*. Available at: https://uk.pcmag.com/wearables/150162/the-best-smart-glasses-for-2024 (Accessed: 25 October 2024).

Hatala, M. and Wakkary, R. (2005) 'Ontology-Based User Modeling in an Augmented Audio Reality System for Museums', *User Modeling and User-Adapted Interaction*, 15 (3), pp. 339–380. Available at: https://doi.org/10.1007/s11257-005-2304-5.

Herre, J. and Disch, S. (2023) 'MPEG-I Immersive Audio – Reference Model for the Virtual/Augmented Reality Audio Standard', *Journal of the Audio Engineering Society*, 71 (5), pp. 229–240. Available at: https://doi.org/10.17743/jaes.2022.0074.

Herre, J. et al. (2024) 'MPEG-I Immersive Audio: A Versatile and Efficient Representation of VR/AR Audio Beyond Point Source Rendering', in. 2024 AES 6th International Conference on Audio for Games, Tokyo, Japan: Audio Engineering Society. Available at: https://aes2.org/publications/elibrary-page/?id=22400.

HTC (2023) 'VR Trackers and Virtual Reality Tracking Explained – VR 101: Part III'. Available at: https://blog.vive.com/us/tracking-in-virtual-reality-and-beyond-vr-101-part-iii/ (Accessed: 27 November 2024).

Härmä, A. et al. (2004) 'Augmented Reality Audio for Mobile and Wearable Appliances', *Journal of the Audio Engineering Society*, 52 (6), p. 22.

Italiano, L. et al. (2024) 'A Tutorial on 5G Positioning', *IEEE Communications Surveys & Tutorials*, pp. 1–1. Available at: https://doi.org/10.1109/comst.2024.3449031.

IVAS Codec Public Collaboration Contributors (2023) 'IVAS Codec for the NG 3GPP Voice and Audio Services', November, pp. 18–19.

Jacuzzi, G. (2018) '"Augmented Audio": An Overview of the Unique Tools and Features Required for Creating AR Audio Experiences', in International Conference on Audio for Virtual and Augmented Reality, Redmond, WA, USA: Audio Engineering Society, p. 8.

Katz, L. (2024) 'Understanding Bluetooth Codecs', *SoundGuys*. Available at: https://www.soundguys.com/understanding-bluetooth-codecs-15352/ (Accessed: 18 October 2024).

Larsson, P. et al. (2010) 'Auditory-Induced Presence in Mixed Reality Environments and Related Technology', in E. Dubois, P. Gray, and L. Nigay (eds) *The Engineering of Mixed Reality Systems*. London: Springer (Human-Computer Interaction Series), pp. 143–163. Available at: https://doi.org/10.1007/978-1-84882-733-2_8.

Lin, Y. and Abdulla, W.H. (2015) 'Audio Watermarking Techniques', in Y. Lin and W.H. Abdulla (eds) *Audio Watermark: A Comprehensive Foundation*

Using MATLAB. Cham: Springer International Publishing, pp. 51–94. Available at: https://doi.org/10.1007/978-3-319-07974-5_3.

Lindau, A. (2009) 'The Perception of System Latency in Dynamic Binaural Synthesis', in 53rd Annual Meeting for Acoustics, DAGARotterdam, pp. 1063–1066. Available at: https://www2.ak.tu-berlin.de/%7eakgroup/ak_pub/2009/Lindau_2009_The_Perception_of_System_Latency_in_Dynamic_Binaural_Synthesis.pdf (Accessed: 18 October 2024).

Lombard, M. and Ditton, T. (1997) 'At the Heart of It All: The Concept of Presence', *Journal of Computer-Mediated Communication*, 3(JCMC321). Available at: https://doi.org/10.1111/j.1083-6101.1997.tb00072.x.

Marvelmind (2023) 'Architecture Comparison'. Available at: https://marvelmind.com/pics/architectures_comparison.pdf (Accessed: 26 November 2024).

Mazuryk, T. and Gervautz, M. (1999) 'Virtual Reality – History, Applications, Technology and Future', *ResearchGate* [Preprint]. Available at: https://www.researchgate.net/publication/2617390_Virtual_Reality_-_History_Applications_Technology_and_Future (Accessed: 10 March 2019).

McGill, M. et al. (2020) 'Acoustic Transparency and the Changing Soundscape of Auditory Mixed Reality', in *Proceedings of the 2020 CHI Conference on Human Factors in Computing Systems*. Honolulu, Hawaii, USA: ACM, pp. 1–16. Available at: https://doi.org/10.1145/3313831.3376702.

Meyer-Kahlen, N. et al. (2023) 'Measuring Motion-to-Sound Latency in Virtual Acoustic Rendering Systems', *AES: Journal of the Audio Engineering Society*, 71 (6), pp. 390–398. Available at: https://doi.org/10.17743/jaes.2022.0089.

Meyer-Kahlen, N. (2024) *Transfer-Plausible Acoustics for Augmented Reality.* PhD thesis. Aalto University. Available at: https://urn.fi/URN:ISBN:978-952-64-1913-8.

Mülleder, A. et al. (2023) 'Do-it-Yourself Headphones and Development Platform for Augmented-Reality Audio', in Audio Engineering Society Conference: AES 2023 International Conference on Spatial and Immersive Audio, Audio Engineering Society. Available at: https://www.aes.org/e-lib/browse.cfm?elib=22188 (Accessed: 8 September 2023).

Naphtali, D. and Rodkin, R. (2019) 'Audio Augmented Reality for Interactive Soundwalks, Sound Art and Music Delivery', in *Foundations in Sound Design for Interactive Media*. Routledge.

Nishimura, T. et al. (2023) 'Cartilage Conduction Hearing Aids in Clinical Practice', *Audiology Research*, 13 (4), pp. 506–515. Available at: https://doi.org/10.3390/audiolres13040045.

Orsholits, A. et al. (2024) 'PLATONE: An Immersive Geospatial Audio Spatialization Platform', in The 2nd Annual IEEE International Conference on Metaverse Computing, Networking, and Applications (IEEE MetaCom), Hong Kong.

Roginska, A. (2018) 'Binaural Audio Through Headphones', in A. Roginska and P. Geluso (eds) *Immersive Sound: The Art and Science of Binaural and Multi-Channel Audio*. Routledge.

Rovithis, E. et al. (2019) 'Audio Legends: Investigating Sonic Interaction in an Augmented Reality Audio Game', *Multimodal Technologies and Interaction*, 3 (4), p. 73. Available at: https://doi.org/10.3390/mti3040073.

Schmalstieg, D. and Hollerer, T. (2016) *Augmented Reality: Principles and Practice*. Addison-Wesley Professional.

Sunder, K. (2022) 'Binaural Audio Engineering', in J. Paterson and H. Lee (eds) *3D Audio*. New York, NY: Routledge (Perspectives on music production), pp. 130–159.

Supper, B. (2021) 'Supperware Head Tracker 1 Specifications'. Available at: https://supperware.net/downloads/head-tracker/specs.pdf (Accessed: 18 October 2024).

Tran, P.K., Amrein, B.E. and Letowski, T.R. (2009) 'Audio Helmet-Mounted Displays', in C.E. Rash et al. (eds) *Helmet-Mounted Displays: Sensation, Perception and Cognition Issues*. Fort Rucker, Alabama: US Army Aeromedical Research Laboratory, pp. 175–234. Available at: http://doi.apa.org/get-pe-doi.cfm?doi=10.1037/e614362011-006 (Accessed: 10 October 2024).

Ungi, T., Lasso, A. and Fichtinger, G. (2015) 'Tracked Ultrasound in Navigated Spine Interventions', in *Lecture Notes in Computational Vision and Biomechanics*, pp. 469–494. Available at: https://doi.org/10.1007/978-3-319-12508-4_15.

Vazquez-Alvarez, Y. et al. (2016) 'Designing Interactions with Multilevel Auditory Displays in Mobile Audio-Augmented Reality', *ACM Transactions on Computer-Human Interaction*, 23 (1), pp. 1–30. Available at: https://doi.org/10.1145/2829944.

Vertucci, R. et al. (2023) 'History of Augmented Reality', in A.Y.C. Nee and S.K. Ong (eds) *Springer Handbook of Augmented Reality*. Cham, Switzerland: Springer (Springer Handbooks).

Wu, Y., Tang, F. and Li, H. (2018) 'Image-Based Camera Localization: An Overview', *Visual Computing for Industry, Biomedicine, and Art*, 1 (1), p. 8. Available at: https://doi.org/10.1186/s42492-018-0008-z.

Yang, J., Barde, A. and Billinghurst, M. (2022) 'Audio Augmented Reality: A Systematic Review of Technologies, Applications, and Future Research Directions', *Journal of the Audio Engineering Society*, 70 (10), pp. 788–809.

Yang, Z. et al. (2020) 'Ear-AR: Indoor Acoustic Augmented Reality on Earphones', in *Proceedings of the 26th Annual International Conference on Mobile Computing and Networking. MobiCom '20*, London, UK: ACM, pp. 1–14. Available at: https://doi.org/10.1145/3372224.3419213.

8   Narrative design considerations

A storyteller never faces a situation where they can pick any medium that best conveys the story, and within that medium, any approach, technology and venue imaginable. The choices are always dictated by the author's skills and interests, the stakeholders' needs and budget, and the priorities of the producer and funder. But let us still assume the storyteller has chosen to use AAR to convey their story. So, where to start? This chapter briefly discusses several narrative design considerations. While practical production-related issues have been touched lightly upon in previous sections and will continue to appear in some later discussions, they are not addressed in depth due to space constraints.

8.1 What is narrative?

A story, or narrative, is 'a chain of events in a cause–effect relationship occurring in time and space' (Bordwell and Thompson, 1997, p. 90). A story is situated and interpreted in a specific discourse context, and it typically conveys the experience of 'what it is like' to be living through the events in the story world (Herman, 2009). Hence, the storyteller's task is not only to explain the progress of events and educate, but to create a world around the events and characters and immerse the audience into that story world (McErlean, 2018).

The noun 'narrative' is commonly used synonymously with 'story'. Yet, narrative also carries its own distinctive meaning as a version of story events, often to create a manifestation or drive an agenda (Halverson, 2011). The idea of presenting a new, alternative view, or *narrative*, to the prominent matters is arguably one of the key concepts of augmented reality. For simplicity, however, in this book, the terms story and narrative are mostly treated as interchangeable since the focus will be more on the way stories and narratives are told, the narrative techniques.

DOI: 10.4324/9781003627289-8

8.2 Narrative potential

As we know, AAR holds huge potential as a narrative medium. One significant advantage of AAR over visual AR is that, already now, the technology allows the creation of fully plausible illusions of virtual objects coexisting with the real environment. This together with the other characteristics of AAR open up several fascinating narrative possibilities, some of which are suggested next.

Familiar environment as a narrative space. AAR enables stories that are situated in the user's own environment, turning the familiar surroundings into a narrative space. *Zombies, Run!* is an example of this, infesting everyone's neighbourhood with zombies, potentially creating a strong emotional reaction. The challenge is, of course, how to individualise the content so that it does not feel too generic. Genres operating at a more primitive cognitive level such as horror can potentially work in this regard, whereas site-adaptive narrative structures may be challenging to design.

Personal and collective stories tied to a place. Hearing is often considered as a personal and intimate sense. Personal stories tied to a particular place and heard spatially with kinaesthetic interaction can be potentially very powerful, *Maison Gainsbourg* with Charlotte's recollections is one such example. Geolocated audio platforms also show potential in this regard, enabling sharing of users' own stories and offering authenticity and social glue to the narratives. Collective storytelling, on the other hand, could uncover shared narratives of a place, possibly enriched with real-time contributions from participants.

Hidden narratives. AAR can reveal 'hidden' narratives of a certain place, offering an alternative perspective to a prominent narrative, exposing ethical, social, and environmental themes. With headphone-based applications, these narrative layers can be hidden from other people and experienced without disruption to the existing environment and surrounding people. Chapter 9.7 discusses a narrative technique called 'alternative match' with some examples of the use of this concept.

Collective and personal. Related to hidden and alternative narratives, one possibility enabled by headphone-based AAR is to offer participants a shared story, but individualised for each listener. Different language options are an obvious possibility, as well as customisation based on age, hearing impairments, or cognitive challenges. However, the customisation can be taken to the narrative level, letting each participant hear the same story in different temporal order or from different characters' perspectives like in *The Reign Union*.

Augmented theatre. While some theatres offer translation services through headphones, AAR could make the experiences more immersive by spatialising the translated voices, attached to the actors on stage. Narratively, this would offer interesting possibilities for personalising some of the dialogue as described above. Additionally, virtual sound effects and ambiences could be delivered through the AAR system, which is potentially useful for some immersive effects that would be challenging to realise with physical loudspeakers around the audience. Some theatre performances such as *Anna* at the National Theatre, UK, have utilised binaural telepresence, letting the audience hear the stage events through headphones. Advanced AAR technologies could be used to take these kinds of concepts further.

Bringing back extinct sounds and evoking the future. Museums have long used virtual audio to bring back sounds that cannot be heard anymore. The *Nature Soundscapes* project introduced sounds of extinct species in a forest through loudspeakers, and with mobile AAR systems, the concept could be scaled to wider audiences. Similarly, AAR could suggest how the world around us would sound in the future, with or without human influence.

Emergent narratives. AAR enables environment-driven narratives that react and adapt to environmental factors such as weather (e.g., Naphtali and Rodkin, 2019). With the help of AI, generative stories could also totally emerge from the reality around the user, receiving inputs from both the environment and people within it.

8.3 Technical platform

As we have learned throughout this book, there are—in theory—an unlimited number of different possibilities to realise AAR. When it comes to reality, the practical choices are, however, often limited and already largely defined by the intended environment and location-dependency, for instance, whether the application will be ubiquitous or site-specific.

Direct augmentation is a clear option if that fits the needs. Usually, simple interaction such as proximity sensing is relatively easy to realise, although making the system reliable for public settings may take effort. For mobile experiences in outdoor environments, geolocated apps and SDKs offer easy authoring, steady performance and global compatibility with a large number of supported end devices. However, the trade-off is limited artistic and interaction tools, unless tailored with the developers of these systems.

For headphone-based indoor experiences, to the knowledge of the author, there are no comprehensive, openly available, easy-to-use, and reliable platforms for content creation and distribution. Some geolocated platforms such as *Echoes* support iBeacons or other proximity sensors, but their positional resolution may not be enough for most applications. To get full 6DoF, one way is to approach the owners of proprietary systems with a project proposal—or money. Otherwise, do-it-yourself solutions may be the only option through building an own system with computers or mobile devices, game engines or other development environments, virtual audio plugins, and experimenting with various tracking systems. With such an approach, high quality can be obtained, while the amount of work may turn out to be unfeasible at least for commercial endeavours.

While all AAR platforms are still somewhat experimental, CTC, WFS, and surround systems fall even further into that category when it comes to AAR—or are simply out of reach for most creators and audiences. Still, creating AAR content for them would be highly encouraged!

8.4 Which environment?

Another key consideration—tightly connected to the technical platform—is whether the story should be site-specific, transferable to multiple sites, or mobile and available everywhere. Writing a story tied to a specific place is likely to result in a cohesive narrative that optimally utilises the real environment, its history, future, and inhabitants. However, there are accessibility issues: not all audiences can easily access site-specific narratives due to geographic location, physical disabilities, or socioeconomic barriers. This can limit the diversity of the audience and exclude those who may benefit from the experience. Further, the need for audiences to travel to the site can contribute to environmental pollution, particularly if the site is not easily reachable by public transportation.

Transferable and ubiquitous experiences would solve the accessibility problem, but pose a challenge for narrative design: how to make meaningful stories that adapt and ground the story to any environment? One solution is to hand pick the sites where the experience runs, for instance, several parks around the world (e.g., *The Planets*), or limit the possible settings, for instance, make the story take place in a domestic environment (e.g., *Horror-Fi Me*), a context that is widely familiar and capable of evoking universal emotional responses. The risk still remains that the experience stays more generic and connections to the reality remain superficial. Some genres tolerate that better, e.g., horror and suspense, which operate at more primitive emotional levels and may be more easily transferable to different environments.

More specifically, the experience can be *location-driven*, where the content is triggered based on the user's location within the environment, or *environment-driven*, where information from the user's surroundings determine how the virtual sounds behave, or a combination of these two (Naphtali and Rodkin, 2019). A location-driven experience would typically be built on a map or floor plan where virtual sounds are placed. From the narrative point of view, this paradigm allows precise spatial design where the location of objects and their sounds are mostly known beforehand.

Environment-driven applications would be *context-aware*, requiring a much more dynamic design approach as well as different ubiquitous tracking technology. An environment-driven system can utilise computer vision, LiDAR, microphones, and other sensors to recognise features and objects in the surroundings. The application can be programmed to look for certain types of objects, and when spotting one of that kind, attaching a specific sound onto it. If the object moves, the attached sound would move, too. For example, the AAR audio guide at the *Vienna Augarten* porcelain museum uses an iPhone-based system that recognises each porcelain exhibit by its look and spatially attaches the corresponding voice narration to it (Aichinger, 2024). When, at one point during the exhibition, one of the porcelain items was moved from its original location to between two other exhibits, the system obediently re-aligned the virtual sound to the new position. However, as the sounds were equipped with trigger zones, they now overlapped with each other causing the narrations to behave erratically. Thus, designing and controlling an environment-driven system would take some extra effort.

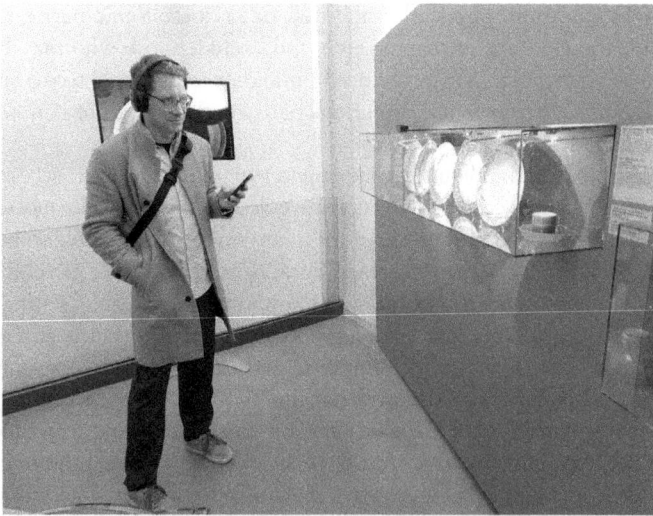

Figure 8.1 A participant in the *Vienna Augarten* experience
Source: Photograph by the author

8.5 Focus

There is likely an infinite number of narrative approaches to an AAR experience. Some of the example applications in Chapter 5 were categorised based on the focus of the augmentations: *AAR in exhibitions, reinforced environments*, and *reskinned environments*. In exhibitions, it is usually the exhibits and the thematic scope of the exhibition that are augmented, while the real environment, such as the venue itself, is typically ignored to some extent, *Dimensions of Sound* serving as an example. The other two categories draw from Robert Azuma's (2015) concepts of *reinforcing, remembering*, and *reskinning*. They are approaches to AR where both the real and virtual are essential to the experience, and can arguably be adapted to AAR, too, as design paradigms.

In reinforcing, the experience is based on a real environment, object or person that is compelling by itself, and the role of virtual sound is to complement that. Many site-specific narrative AAR experiences use this approach, including applications in museums or cultural heritage sites that revive historical objects or environments through sounds. As an example, *Audio Legends* brings the history of Corfu to life through an AAR game. Whilst this approach typically reinforces real-world items and events, virtual sound design can also fuse new and contradictory meanings, if desired.

Remembering, in turn, evokes stories connected to specific locations and retells them where the events originally occurred. The goal is that merging these memories with the actual site will create a more impactful experience than either the location or the virtual content could offer on its own. Audio walks typically utilise this approach, Graeme Miller's *Linked* being a good example. *Maison Gainsbourg* is another example, but it also combines remembering with reinforcing, overlaying powerful memories in the rooms of an already extremely compelling home of the Gainsbourg family.

The reskinning approach has two variations. In the first, the environment can be arbitrary which is then 'recharacterised' by virtual content. The story is placed within the existing environment by utilising surrounding objects such as buildings, vehicles, and people as narrative elements through augmentations. *Zombies, Run!* would be one example, reskinning the familiar neighbourhood as zombie-infested no-man's land. The reskinning approach benefits from a context-aware AAR system making it truly mobile and ubiquitous. In the second variations, the whole environment can be an artificially constructed, a set piece. For example, *The Sound of Things* is based on items that are carefully arranged on a table, each object being real but, at the same time, artificially organised. *The Reign Union*, on the other hand, combines this approach with reinforcing, using a real location as the anchor for the story, but relying on sonic augmentations and set decorations to convey narratives beyond the real space.

A reskinned or entirely artificial environment could form the basis of a 'hybrid reality' approach, as suggested by Schraffenberger (2018), where the physical environment or object is intentionally incomplete and becomes whole only through the addition of virtual content—like a karaoke track that lacks vocals, which are supplied by the singer.

8.6 Compromises

Unless unlimited resources are available, compromises in AAR design are inevitable. High-quality virtual audio is often sacrificed for engaging interactivity, and sometimes both are traded for reliability, low maintenance, and scalability—the next user should get as good an experience as the previous one. In museums, for example, if visitors can trust the technology to work reliably, they will also be more likely to engage with it and have a positive experience (Chen et al., 2024). Not many museums have extra staff to solve problems, and buggy software is quickly uninstalled and forgotten. And even so, any interruption breaks the flow of the experience.

If the production has enough resources to develop natural interaction for the experience and virtual sounds that are at least *contextually* congruent with the chosen genre and environment, the experience seems to tolerate less authentic virtual acoustic rendering while still maintaining immersion (Aichinger, 2024; Armbruster, 2024). Cummings and Bailenson (2016, p. 293) cite Julian Hochberg who suggested that 'Perfect physical fidelity is impossible and would not be of psychological interest if achieved, but perfect functional fidelity is completely achievable and is of considerable psychological interest'. Creating interactive content for AAR is no trivial task, however: beyond the complex design process, it requires recording multiple dialogue or voice line variations, extensive editing, programming, testing, and numerous other work phases—each a significant effort in its own right.

On the other hand, a gripping story told through near-realistic virtual sounds may not require any interaction at all (e.g., *Séance*). So, while resources may limit the ability to include every last feature, a well-curated experience can still fully engage participants and lead them to suspend their disbelief. Further, as is the case with new media, the fascinating medium itself may sometimes draw too much attention away from the content (Green, 2021; Krieger et al., 2023). This phenomenon might occur in the 'uncanny valley', where the technology mimics realism impressively but not convincingly enough for the audience to become fully immersed and accept the illusion of nonmediation.

8.7 Listener's role

In narrative experiences, the listener's role—whether observer, passive participant, or active actor—shapes the interactive demands and technical complexity of the experience. Here are some example roles:

Ghost. The listener is a silent observer, free to move but often unnoticed by other characters and without impact on the story. However, the listener's movements may trigger events in the scene. E.g., *The Sound of Things* and *Sonic Traces Heldenplatz*.

Passive central character. The listener becomes a central figure in the story, yet remains passive, following a directed, immersive narrative with no need for active input. E.g., *Séance*.

Passive companion/visitor. The listener is invited to explore as a visitor, following pre-defined paths or encountering scenes curated for them. Here, the interactivity is indirect, with the listener following guidance through soundscapes or narratives that create a sense of place but do not require active decision-making. This role still leans on script-driven audio but requires structured environmental cues that signal exploration. E.g., *Maison Gainsbourg*.

Active actor. The listener actively shapes the narrative through their choices and movements, effectively becoming a character within the story. This requires complex interactivity, with the user's actions driving dynamic responses and narrative paths. E.g., *The Reign Union*.

There are many other roles to consider, but the key point is that the interactive requirements of each role dictate the types of input needed. In more passive roles, inputs can be limited to location tracking or simple presence detection. Active roles, however, require a richer set of inputs, perhaps 6DoF movement, gestures, and voice.

In multi-user AAR, the narrative may need to adapt to interactions between users. This naturally introduces additional layers of complexity to storytelling. An ever more complex but fascinating concept would be a common but distributed experience shared across multiple users in their remote locations, something envisioned at the beginning of this book.

8.8 Secondary user interface

As discussed in Chapter 2, the primary user interfaces of typical AAR applications are the auditory display and movement. However, there is often a need for additional input interfaces such as volume controls, scene selection buttons, or voice recognition for speech commands. Additionally, beyond audio, certain applications may benefit from secondary output methods, such as graphical data on a mobile phone screen. Especially if the experience is accessible by the user alone, through downloading a mobile app or picking up headphones in a self-service principle, an understandable and functional UI dictates whether the user understands the context and feels comfortable in proceeding to the experience. As Naphtali and Rodkin (2019, p. 312) point out, 'If users become frustrated with merely accessing and managing the piece's content, they simply will turn it off and not use it.'

Audio can also serve as a secondary interface: besides the environment-embedded virtual audio content, a head-locked audio layer can be used simultaneously for system-level messages, and this may even be a good idea to reduce cognitive load (Vazquez-Alvarez et al., 2016).

The potential problem with secondary interfaces is that they may distract the immersion or 'flow' (Brandenburg and Sloma, 2024). Operating a touchscreen while being engaged in a narrative may remind of the mediation process in place (see Chapter 3.2.). Besides narrative applications, many ubiquitous systems attempt to remain invisible to the user, too.

Table 8.1 Examples of secondary interfaces of AAR

Domain	Input	Output
Auditory	Speech recognition	System messages
Visual	Gesture recognition Facial tracking	Text and graphics
Physical/tactile	Touchscreen Physical knobs, sliders Tapping on the side of device	Vibration

Intuitive interactions such as tapping on the side of smart glasses, voice commands, or hand gestures, like pinching with thumb and index finger, are getting common, and when users get used to them, they probably get to be used 'automatically' without breaking the immersion.

One approach is to harness artificial intelligence to make decisions for the user, thus minimising the need for secondary interfaces, but the obvious risk is wrong interpretations by the machine. The user interface can also be disabled or muted when it is likely not needed. For instance, a novel AAR exhibition system could provide thorough instructions in the beginning and require volume and other adjustments by the user, after which these secondary interfaces could be disabled. Some video games, such as *Journey* from 2012, manage to teach gameplay mechanics seamlessly by integrating them into the narrative and environment. These games allow players to learn through exploration and interaction, rather than through explicit tutorials or instructions. Similar approaches in AAR, too, would immediately capture the user's attention from the very start of the experience.

8.9 Accessibility considerations

Since AAR is well-suited for individuals with reduced vision, and many potential users may also have hearing impairments, it is important to consider a few key design principles to address these specific needs. First of all, references to visual objects should obviously be avoided, such as 'Walk towards the yellow building'. Instead, more accessible cues could be used, such as audible or tangible reference points like 'Turn towards my voice', or instructions relative to the user's current pose: 'Turn left. A bit more still... Perfect!'

Further, care should be taken with audio levels. Many blind individuals use reverberation cues and environmental sounds to comprehend the surroundings. However, even with a transparent audio system, too loud virtual sounds can easily mask the real-world sounds. As an example, in one scene of *The Reign Union*, music gets spatialised around the user as a three-dimensional, acousmatic orchestra. The dialogue encourages the user to start dancing, but some of the blind participants decided to stop completely as they could not hear their real-world surroundings any more, feeling uncomfortable to move. Addressing this problem by lowering the volume of the virtual sounds would probably undermine the power and magic of the scene. Hence, a better solution could be to give the user the option to stay still and enjoy the music.

For individuals with hearing difficulties, using realistic sound levels may not always be possible. Real-world sounds are often very quiet. In narrative AAR experiences, scenarios like eavesdropping on a conversation through a ventilation shaft or behind a door provide unique opportunities to incorporate obstructed sounds (see Chapter 9.2). Practical experience shows that such virtual sounds are often easier to render plausibly than those attached to exposed objects—a principle that also applies to distant sounds.

However, for users with reduced hearing, these sounds may become inaudible or so faint that attempting to discern dialogue or other details becomes stressful. Allowing users to raise the volume could improve intelligibility but risks breaking the acoustic illusion. Hence, some compromises must probably be made, prioritising intelligibility over realism.

For users with hearing aids compatible with AAR auditory displays, such as headphones, these issues may be mitigated. Ideally, the virtual audio scene would closely mimic the real acoustic environment, enabling the hearing aid to enhance sound—either automatically or through manual adjustments—without disrupting the immersive experience.

8.10 Level of realism

The level of realism in AAR is not just an aesthetic choice. It can affect how users understand the narrative: which sonic elements they recognise as being part of the story. In a traditional theatre play, film, or video game, the participant can, at any moment, look around and distinguish the story world from the real world despite how realistically the characters, settings and sounds are portrayed. With AR, immersive theatre, live role-playing games, and some experimental video games, that line is blurred, since everything around the user can be a part of the story. With AAR, it is possible to render virtual content so plausible that one cannot distinguish it from the real even when making a conscious effort to do so.

An inability to know which elements belong to the story and which ones to the real world can evoke a sense of uncertainty and create confusion (see Härmä et al., 2003; Dam et al., 2024). In controlled environments, such as the DARKFIELD shows in shipping containers, the realism dial can be turned all the way up, as each audience member knows they are entering a fully immersive experience where every sound might—or might not—play a role in the story. However, in experiences located in public spaces or surrounded with other people, using near-realistic virtual sounds may pose a risk in case the user does not understand them to be part of the narrative and starts to behave unpredictably. As the legend goes, during the broadcast of *The War of the Worlds* in 1938, which was an episode of Orson Welles' radio series *The Mercury Theatre on the Air*, some people took the Martian invasion for real and panicked. In AAR, the same could happen, for instance, in a museum submarine where a virtual fire alarm went off, or if the user heard an act of violence happening behind a door, causing them to call the police. In Dafna Naphtali's geolocated audio walk *Walkie Talkie Dream Angles*, located at the Washington Square Park in New York, the makers did not want to use the sounds of fire truck sirens as they would have masked the real emergency vehicles (Naphtali and Rodkin, 2019).

In this world of deepfake videos and voice cloning, 'audiorealistic' sounds may also raise concerns about digital deception: the ability of AAR to create virtual sounds that are difficult to distinguish from physical ones could make users more susceptible to deception (Turner, 2022).

Using stylised sound design likely reduces the risk of the aforementioned problems: the experience may entail an engaging and believable story, the possibility to move within the scene, and affordances available to the player, all immersing the participant and letting them forget that they are following a story (see Wirth et al., 2007; Järvinen, 2017; Green, 2021). However, the fact that the virtual and real soundscapes are still distinguishable from each other, would provide a possibility for a 'sanity check' if doubts arise.

It may well also be that plausible virtual illusions are not technically possible to realise even if desired. That does not mean that the experience is doomed to failure as there are many other factors contributing to a captivating experience, like the elements described in the previous paragraph and also discussed in Chapter 3.2.

8.11 Conclusion

The decision to jump into an AAR production is already a leap into the unknown, and moving forward from there often relies on first-hand exploration and learning by doing. The general design considerations outlined in this chapter are intended to help navigate the complexities of AAR design. However, the practical aspects of production, which are equally crucial for the success of an AAR experience, remain a topic for further exploration. While perfect auditory illusions can be enchanting and enable true mixed reality experiences, they are not always feasible— or even necessary—for many applications. This realisation likely provides reassurance to practitioners, broadening the scope of what can be achieved. It also underscores the importance of balancing technical precision with creative adaptability.

References

Aichinger, T. (2024) 'Conversation with Matias Harju, March 6'.

Armbruster, S. (2024) 'Conversation with Matias Harju, March 7'.

Azuma, R. (2015) 'Location-Based Mixed and Augmented Reality Storytelling', in B. Woodword (ed.) *Fundamentals of Wearable Computers and Augmented Reality*, 2nd edn. CRC Press, pp. 259–276.

Brandenburg, K. and Sloma, U. (2024) 'Conversation with Matias Harju, May 23'.

Bordwell, D. and Thompson, K. (1997) *Film Art: An Introduction*. 5th edn. The McGraw-Hill Companies.

Chen, Y. et al. (2024) 'Why People Use Augmented Reality in Heritage Museums: A Socio-Technical Perspective', *Heritage Science*, 12 (1), p. 108. Available at: https://doi.org/10.1186/s40494-024-01217-1.

Cummings, J.J. and Bailenson, J.N. (2016) 'How Immersive Is Enough? A Meta-Analysis of the Effect of Immersive Technology on User Presence', *Media Psychology*, 19 (2), pp. 272–309. Available at: https://doi.org/10.1080/15213269.2015.1015740.

Dam, A. et al. (2024) 'Taxonomy and Definition of Audio Augmented Reality (AAR): A Grounded Theory Study', *International Journal of Human-Computer Studies*, 182, p. 103179. Available at: https://doi.org/10.1016/j.ijhcs.2023.103179.

Green, M.C. (2021) 'Transportation into Narrative Worlds', in L.B. Frank and P. Falzone (eds) *Entertainment-Education Behind the Scenes*. Palgrave Macmillan, Cham, pp. 87–101. Available at: https://doi.org/10.1007/978-3-030-63614-2_6.

Halverson, J. (2011) 'Why Story is Not Narrative', CSC Center for Strategic Communication, 8 December. Available at: http://csc.asu.edu/2011/12/08/why-story-is-not-narrative/ (Accessed: 19 April 2021).

Herman, D. (2009) *Basic Elements of Narrative*. Hoboken, UK: John Wiley & Sons, Incorporated. Available at: http://ebookcentral.proquest.com/lib/aalto-ebooks/detail.action?docID=437514 (Accessed: 17 April 2021).

Härmä, A. et al. (2003) 'Techniques and Applications of Wearable Augmented Reality Audio', in. *Audio Engineering Society Convention* 114, Audio Engineering Society. Available at: http://www.aes.org/e-lib/browse.cfm?elib=12495.

Järvinen, A. (2017) 'Design for Presence in VR, Part 2: Towards An Applied Model', *Medium*. Available at: https://virtualrealitypop.com/designing-for-presence-in-vr-part-2-towards-an-applied-model-2784bf16a01 (Accessed: 29 November 2024).

Krieger, E. et al. (2023) 'Youth and Augmented Reality', in A.Y.C. Nee and S.K. Ong (eds) *Springer Handbook of Augmented Reality*. Cham, Switzerland: Springer (Springer Handbooks), pp. 709–741.

McErlean, K. (2018) *Interactive Narratives and Transmedia Storytelling: Creating Immersive Stories Across New Media Platforms*. 1st edn. Milton: Routledge. Available at: https://doi.org/10.4324/9781315637570.

Naphtali, D. and Rodkin, R. (2019) 'Audio Augmented Reality for Interactive Soundwalks, Sound Art and Music Delivery', in *Foundations in Sound Design for Interactive Media*. Routledge.

Schraffenberger, H. (2018) *Arguably Augmented Reality: Relationships Between the Virtual and the Real*. PhD thesis. Universiteit Leiden. Available at: https://www.creativecode.org/wp-content/uploads/Thesis/thesis-print.pdf (Accessed: 14 November 2024).

Turner, C. (2022) 'Augmented Reality, Augmented Epistemology, and the Real-World Web', *Philosophy & Technology*, 35 (1), p. 19. Available at: https://doi.org/10.1007/s13347-022-00496-5.

Vazquez-Alvarez, Y. et al. (2016) 'Designing Interactions with Multilevel Auditory Displays in Mobile Audio-Augmented Reality', *ACM Transactions on Computer-Human Interaction*, 23 (1), pp. 1–30. Available at: https://doi.org/10.1145/2829944.

Wirth, W. et al. (2007) 'A Process Model of the Formation of Spatial Presence Experiences', *Media Psychology* [Preprint]. Available at: https://doi.org/10.1080/15213260701283079.

9 Narrative techniques and concepts

AAR is a new medium with huge narrative potential. Although artificial sounds have been merged with reality for decades to tell stories related to specific places or objects, content creators often work in isolation, unaware of others' efforts in the field. Many experiences incorporating AAR elements remain unlabelled as such, making it challenging to learn from or build upon them. While this book seeks to bring these applications together under a shared framework, it also identifies the unique narrative techniques employed in these experiences. By recognising common tools and practices within the AAR community, the effort to create content could become slightly easier. This would allow creators to focus more on exploring and developing new ideas.

In this chapter, several narrative techniques and concepts characteristic to AAR are presented (see Table 9.1). They mainly rely on the use of spatial virtual audio, interplay between real and virtual, and interactivity based on the user's location and movements. Most of the techniques have been identified by the author within the Full-AAR project (see Chapter 5.4). This work draws upon firsthand experience with several AAR projects, interviews with AAR content creators, research literature, and insight gained from developing *The Reign Union* as part of Full-AAR.

The techniques presented in this chapter are not exclusive to AAR. They borrow from and may be applicable to other multimodal mediums, such as cinema, video games, visual-based AR, and VR. Similarly, many narrative and sound design approaches used in other contexts would likely be effective in AAR as well. That said, the focus here has been to identify and elaborate on techniques that are particularly *characteristic* of AAR, even if they overlap with other mediums. Some potentially interesting techniques have been excluded to maintain focus, especially if they are widely employed in other interactive and immersive formats. For example, narrative sequencing based on interaction has been left with little attention in here for this reason, together with a number of

DOI: 10.4324/9781003627289-9

other relevant topics. Drawing the line has not always been easy. Undoubtedly, many AAR-specific techniques also remain unidentified.

Narrative techniques here refer to methods to accomplish storytelling tasks: conveying information and emotions, developing the story, and making it more engaging and complete. In established art forms, the techniques and their effects, the narrative codes, are familiar to the audience. For example, spectators of conventional cinema have no difficulties in following stories presented through the use of montage (Kuhn, 1985), even though the effect of this technique is quite removed from everyday human experience. In younger mediums, however, narrative techniques and codes are still strongly evolving, which is arguably true for XR experiences in general and AAR in particular. This requires a learning process for both the audience and the creators.

In AAR, a key narrative consideration is the relationship between virtual sounds and their real-world counterparts. In other words, the narrative can be significantly influenced by whether sounds are attached to physical objects, and whether they 'match' or contrast with those objects or the environment. These themes underpin the concepts explored in this chapter. While some of the techniques are platform-agnostic, many of them require a binaural audio display with 6DoF or at least XY-Yaw motion to be fully effective.

Most of the narrative techniques listed in this chapter are demonstrated in videos with binaural audio. Links to the videos can be found on the book's eResource on the Routledge website.

9.1 Locative audio

Locative audio refers to acousmatic sounds that are triggered and manipulated based on the user's location. (For acousmatic sounds, see Chapter 2.11.) The technique mainly applies to mobile and wearable AAR, and the audio is typically soundscapes, voice, and music. Locative audio is the de facto paradigm in most geolocated audio applications. This is largely because head tracking is often difficult to implement or not accessible to all users, limiting the ability to attach virtual sounds to objects. In contrast, location tracking is easily achieved with satellite positioning outdoors and is relatively straightforward indoors using proximity sensors.

Narratively, locative audio naturally focuses on meaningful locations, places, and the user's relationship with them, making the user an active explorer through movement. Typically, audio content is started or faded in when the user enters a 'geofenced' trigger zone, and faded out when

Table 9.1 Some narrative techniques and concepts characteristic of AAR

Concept/Technique	Definition	Example
Locative audio	Acousmatic sounds triggered and manipulated based on user's location	Head-locked forest soundscape starts to play when user enters exhibition space (*Dimensions of Sound*)
Attachment	Virtual sound appears as emanating from a real-world object	Virtual knock on a door (*Horror-Fi Me*)
Object congruence	Fully plausible contextual alignment of an object and its sound	Virtual telephone ringing sound attached to a mute desk phone (Cohen, Aoki, and Koizumi, 1993)
Partial congruence	Contextual alignment of an object and its sound with elements that are slightly tangential	Celebratory sounds emanating from an empty wine glass (*The Sound of Things*)
Incongruence	Contextual mismatch between an object and its sound relative to expectations	Dog sounds emanating from a person (*The Reign Union*)
Addition	Introduction of supplementary sounds that manipulate the perception of a sound-producing object	Creaking cable sounds layered over the whir of a lift to transform the safe elevator into an unsettling place
Alternative match	Use of audio that provides a different yet still congruent perspective on an object	In presence of bad odor, one hears idling car, another leaking gas pipe
Extension	Space beyond the observable field created and suggested by sound	Sound of dropping trash making litter bin appear extremely deep (*The world's deepest bin*)
Revelation	Source of an obstructed sound gets revealed	Source of a metallic drop sound revealed in the next room (*Barque Sigyn*)
Acousmatic sounds	Sounds without physical counterpart	Sounds of a family spatialised in the middle of an empty kitchen (*Maison Gainsbourg*)
Detachment	Sound disengaged from a physical object; becomes acousmatic	Touching a sound-emitting exhibit causes sounds to circle around the user

Concept/Technique	Definition	Example
Attractor sounds	Objects beckoning the user with distinctive sounds	Museum exhibit 'calling' from another room (*LISTEN*)
Spatial scale and offset	Spatial coordination system and scale manipulated in various ways	Hearing the soundscape as if from the ears of another person (*National Theatre: Anna*)
Near field	Utilising sounds very close to head	Sound of a mosquito causing irritation (*The Reign Union*)
Removal and replacement	Removing or replacing sounds in user's perception	Replacing person's speech with a scripted version
Synchronisation	Creating momentary synchronisation points between virtual and real world for enhanced meanings	When wall clock strikes 12, street sounds shift to nature sounds
Acoustic translocation	Real space acoustically transformed to a virtual one	Gallery space sounds like a garden (*The Reign Union*)

they exit. However, user's location can be mapped to any parameter such as audio filters, volume, or why not synthesiser 'knobs'.

Trigger zones can be defined on a map or floor plan during the authoring process. The zones can also be relative, that is, tied to reference points or objects in the environment. Besides full location tracking, proximity sensors is one way to realise that, while context-aware systems with computer vision is another: they recognise visual features in the surroundings—such as museum exhibits—and can trigger audio when the user is at a predefined distance from them (e.g., *Vienna Augarten*).

9.2 Attachment

In contrast to locative audio, which relies on acousmatic sounds, virtual sounds can be spatially aligned with real-world objects, creating the impression that the sound is emanating directly from the object. This approach is here called *attachment*. Using this technique, a photograph on a table can 'talk', and muffled sounds of a party next door can be heard as virtual sounds emanating from the wall.

For informative applications, attached sounds can direct the user's attention to a specific object: waypoints in a navigation app could signal their location (e.g., Albrecht, Väänänen, and Lokki, 2016), while in a fighter pilot's helmet, an incoming missile can be heard as a virtual sound originating from its real-world position, enhancing situational awareness (see Chapter 5.5).

In narrative contexts, attachment may be used to create the illusion of an inanimate object coming to life: in *The Sound of Things*, items on the table emanate sounds, as if imbued with a sense of life. On the other hand, the attached sound itself becomes materialised, integrating concretely into the environment. This can make the listening experience feel more 'direct', familiar, and accessible compared to hearing the same audio through more traditional means without spatialisation (Cliffe, 2024).

While general ambiences and soundscapes help the user to immerse in the narrative at hand, attached sounds on individual objects may invite for interaction, deeper engagement, and eventually knowledge through them (Wakkary et al., 2004). It must be noted, however, that ambiences can also be created through attached sounds, although the function of such sounds would generally be to remain more peripheral.

9.2.1 Exposed and obstructed

If the physical object is within our surroundings and detectable by our senses, it can be considered an *exposed* object. The reason to call it exposed, rather than 'visible', is to cater for situations where visual perception is not available, for example, due to impairment or if the experience happens in darkness.

Virtual sound can also be attached onto an object that is *obstructed*, or hidden, behind a wall for instance, or far away or possibly occluded by fog. Unlike light, sound waves propagate through many obstacles and diffract around corners, so it would be plausible that we can still hear the sound even without directly perceiving its origin.

The concepts of exposed and obstructed are not fully water tight as there may be a situation where an object is, for instance, visually obstructed but can still be sensed through other senses, such as smell. For the scope of this book, these can be considered as special cases without immediate need to rework the terminology. However, as with this and many other aspects of AAR, there is always room for the community to refine and adapt the concepts and terminology to better capture the diverse phenomena.

Exposed objects do not need to be visually perceivable, although that is usually the case. Distance, fog, and lack of light, along with an individual's physiological factors, can prevent visibility. Nothing hinders attaching virtual sounds to real-world objects that are invisible for human eye, but perceivable with other senses. Although use-cases may be rare, one can think of adding a sound to a cloud of perfume, for instance, or a hot spot on the wall caused by a heating pipe. In *Séance*, the virtual audio objects are attached to other participants, the container's walls, and for instance a table that the user touches. None of these can be visually perceived during the performance, but they are obviously noticed before the lights are switched off, and some of them can be sensed in the darkness during the show.

9.2.2 Technical requirements

Attachment requires the system to be capable of rendering the sound so that it appears as coming from its real-world position—or at least its true direction. In practical terms, arguably the simplest way to achieve the effect is to use direct augmentation: hide a loudspeaker inside or behind the object. With binaural and other technologies, however, a more sophisticated system is obviously needed. The requirements for binaural systems actually get quite high, because if the user has any difficulties in perceiving where the sound is coming from, the illusion of attachment easily collapses. As discussed earlier, psychoacoustic, visual and narrative cues help in sound localisation and thus support attachment to a point, but it is not guaranteed that they work every time and with every person similarly. In the worst-case scenario, bad externalisation or drift in head-orientation tracking causes attractor sounds and acoustic warnings to be spatialised incorrectly, thus misleading the user.

Regarding pose tracking, there are some situations where the requirements are less tight. If the virtual sounds encompass the user somewhat uniformly from all directions, like a forest, tracking head orientation may not be that crucial. This limited tracking method is widely used in most of the geolocated audio walk applications. On the other hand, with sounds far away, knowing the user's location may not be that critical. The distant rooster crow sounds more or less the same regardless of whether one listens to it from 95 or 105 metres away, or walking a few metres to the left or right assuming there are no acoustic obstacles coming in between when changing position.

In addition to direct augmentation and binaural systems, attachment can also be achieved technically with at least directional speakers and wave-field synthesis (WFS). However, as discussed in Chapter 2.10, the challenge with WFS is the requirement of multiple loudspeakers that may be difficult to conceal or justify narratively.

In terms of audio production and rendering pipeline, attachment is not exclusive to object-based audio (OBA); similar effects can be realised with scene-based audio pipelines, such as multi-point HOA, and even through head-locked binaural recordings using a dummy head, like in *Séance*. However, a better term for attachment in these cases would probably be 'spatial alignment' as there are no individual sound objects that can be 'attached' or given positional coordinates.

9.2.3 Within-reach and out-of-reach

There are some practical consequences whether the augmented object is *within-reach* or *out-of-reach*. When within-reach and thus inside the 'play area', the user can come close to it and possibly move around it. This sets high requirements for virtual audio rendering; for instance, with binaural systems, accurate and low-latency 6DoF tracking is needed. Far away sounds, objects behind a window, sounds leaking from adjacent rooms, and general ambiences are all out-of-reach—the user cannot reach them. These sounds can be realised with more ease as the spatial accuracy is not that critical. Instead of object-based audio, head-centred Ambisonics can deliver many of out-of-reach sounds.

9.2.4 Cross-modal support

In case there are technical difficulties in aligning the virtual sounds correctly with the real-world objects, the ventriloquism effect, or visual cues, may help, as discussed earlier. Other senses may also come at hand; for instance, if the user can touch the object while hearing its virtual sound coming from a wrong direction, proprioception—the feeling of position in this case—could correct the localisation perception (see Pick, Warren and Hay, 1969). Also, contextual congruence between the object and sound may at least help the user to understand to which object the misaligned sound belongs (Yang et al., 2020).

That said, according to the practical experience, if head tracking in a binaural system is not stable and the virtual sound source keeps erratically moving, the aforementioned cues will not be enough to stabilise the sound image. A quick-fix could be to add smoothing to the tracking data rather than allow constant movements even with the price of reduced responsiveness.

9.2.5 Terminology

While the terms 'attachment' and 'attached sound' are intuitive and rather commonly used to describe this concept, the terminology is not standardised. Some other terms describing the same concept are *spatially registered* (Kaghat and Cubaud, 2010; Tsepapadakis and Gavalas, 2023), *localized* (Härmä et al., 2004; Brandenburg and Sloma, 2024), *anchored* (Väänänen-Vainio-Mattila et al., 2013), and *spatially aligned*. Each term carries a slightly different meaning and is useful whenever it better describes the situation at hand.

9.3 Object congruence

In the beginning of the book (Chapter 2.6), the concepts of plausibility, verisimilitude, and congruence were briefly discussed. Deliberately adjusting the level of congruence between virtual sounds and their real-world counterparts can serve as a narrative technique in AAR. To distinguish this match from other congruences such as audiovisual or room congruence, it is here called *object congruence*. Object congruence, therefore, refers only to the contextual alignment of the object and its sound, not the acoustic properties of the virtual sound.

Object congruence can be used to complement and reinforce the object, although with a concrete and literal match there is also the risk of tone painting and underscoring (Naphtali and Rodkin, 2019) with potentially an uninteresting and boring effect (Stevens, 2009; Fry, 2019).

What is perceived as congruent, of course, varies from person to person and from one experience to another. Cultural background, learned associations, and context are likely to affect this perception (Stevens, 2009). These are the reasons why, in sound design specifically, audio should align with the listener's mental model and expectations rather than how it would actually sound in the real world (Lyons, Gandy and Starner, 2000; Chion, 2019). The genre, or context, has a significant role here. For instance, dogs do not usually talk in real life, while in cartoons—and perhaps in the circus—they do.

Object congruence would require adding, or replacing, sound on an object that *could* produce or *is* producing sound in reality. Examples of objects that could plausibly produce sound are a loudspeaker, running machine, other person, video projection, and any obstructed object that can only be perceived through its sound. In museums, there are often artefacts that used to make sound in the past but no longer do so, like a historical radio receiver. Restoring the sound of such a *silenced* object (Cliffe, 2022) can create object congruence, provided the sound makes the object appear as though it is once again producing that sound.

However, for the sound to be in full congruence with the radio, the device should be switched on, and the frequency dial should be on the correct radio station. The congruence is, however, always highly dependent on the context. For instance, when an old radio from the 1930s is switched on, in many places, you would likely hear only static, as 'longwave' and 'medium wave' stations have become rare. However, in a historical narrative context, hearing a virtual broadcast makes perfect sense. So, as it happens with any narrative medium, the listener here adjusts themselves to the proper context and starts to play along, unless they are very cynical or the experience is somehow badly designed and realised.

The virtual sound might be the original one, such as a recording of an old radio program, or a close approximation, like a re-enacted version. Alternatively, the sound could be entirely fictional or fabricated, as long as it plausibly seems like something the object could produce. For example, during an 'old' radio interview, the guest might start predicting the future, aligning uncannily well with present-day events. Utilising object congruence in this manner represents one of the key narrative and artistic devices of AAR: creating new meaning through an acoustic illusion (see Cliffe, 2022). In other words, the sound alters our perception of the object, adding value through the juxtaposition it creates (Chion, 1994).

Figure 9.1 An old radio receiver in *The Reign Union* with power switched on and augmented with virtual radio broadcast
Source: Photograph by the author

When the illusion works, the user remains uncertain whether the audio-object event is real or virtual. A compelling example can be found in *Séance* and other DARKFIELD productions, which use virtual audience voices. In complete darkness, where the faces of other patrons are invisible, it becomes impossible to determine whether the talking and other sounds are genuinely produced by the people present or artificially created. The DARKFIELD team further enhances this illusion by localising their shows to align with the language and dialect of the region where the performance takes place (Kadel and Rees, 2024).

The congruence illusion also works when applied to people in motion. In *The Reign Union*, the other virtual main character, Mauno, starts to hallucinate and hear whispers of spies everywhere. The user gets to hear the whispers together with him. Now, if another participant enters the same room, a whisper is attached to them and played whenever their head is turned away from the user or is positioned behind the user, ensuring their mouth remains unseen. This avoids any incongruences with asynchrony of lip movements. If the narrative allowed, whistling would create perhaps an even better congruence, since it is more agnostic to the unique timbre of the person's voice.

With obstructed objects, congruence is easy to realise: muffled sounds through walls and doors often appear as surprisingly real. The likely reason is that the sound level is often low, rendering the sounds peripheral and subject to less critical scrutiny. Also, the spatial positioning of such sounds is always somewhat blurry, which forgives any tracking and registration errors in the virtual audio system.

9.4 Partial congruence

Partial object congruence, or *partial congruence* for short, is perhaps the most typical congruence level in narrative AAR applications. At the same time, it is a somewhat vague concept between congruence and incongruence. When partial congruence is applied, some characteristics of the virtual sound align with the object or real environment, while others do not. In such cases, the context and the user's willing suspension of disbelief are crucial factors in determining whether the juxtaposition is considered plausible.

When music and image have partial congruence, it is shown that the spectator's attention draws towards music, away from the image (Stevens, 2009). We can cautiously assume a similar effect occurs in AAR, where a partial mismatch between the virtual sound and the physical object may draw greater attention to the sound component. However, the narrative benefits might outweigh this potential drawback.

Examples of partial congruence in AAR include historical street sounds heard through a curtained window. The listener may not be able to verify whether there are horses and carriages on the street, but unless a historical re-enactment is taking place outside, they can reasonably assume they have not traveled back in time. Similarly, a switched-off radio playing a news broadcast restores sound to a silenced object that could plausibly produce it, even though it is clearly incapable in its current state. In such cases, contextual alignment is rather strong, requiring only a modest suspension of disbelief. By contrast, hearing crackling fire from a pile of unlit logs creates a symbolic augmentation—there is a connection, but it demands a different interpretative mindset from the observer.

Virtual sound can, of course, be attached on a completely *silent* object that has never been able to produce any sound (Cliffe, 2022). For this to work, however, clear narrative congruence is essential. With such, a photograph can easily talk (*The Reign Union*) and an empty wine glass echo the sounds of last night's party (*The Sound of Things*). The result can be symbolic, poetic, and unrealistic, but still plausible if the spatial rendering and internal narrative logic are solid. Since congruence is a continuum, these examples could well be labeled as *incongruence*, too, discussed in the next section.

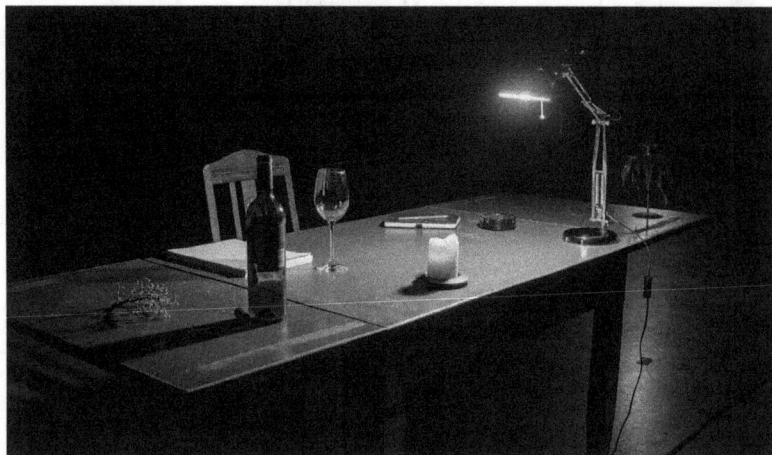

Figure 9.2 Wine glass and other objects emanating sounds in *The Sound of Things*
Source: Photograph by Holger Förterer

In *Nature Soundscapes* (Lawton, Cunningham and Convery, 2020), animal sounds were played back in a forest in central England. In the first scene, the sounds were local and matched the environment. The next scene introduced animals completely out of context, such as bears and monkeys. Finally, human-made sounds were layered over the scene to reflect environmental degradation. While the study's primary focus lay elsewhere, these sounds illustrated varying levels of congruence, with the alien animals representing the furthest deviation from full congruence, while native animals and human-made sounds remained relatively plausible.

Animatronics and humanoid robots are particularly interesting in this regard, as they are inherently silent, artificial, and inanimate, yet designed to appear as the opposite. Despite this intent, their jerky movements and doll-like features reveal their non-human nature. Sounds attached to them are often recordings of real people or animals, even though slightly clunky, mechanical, or robot-like voices might achieve greater congruence and better reflect their level of realism (see Chapter 3.1).

9.5 Incongruence

Incongruence, or mismatch, between expectation and reality is a powerful, yet difficult, narrative tool. As Stevens (2009) suggests, while partial congruency is likely to *enrich* the experience, adding incongruent elements may lead to *emergence* of something new. Hearing a scream each time someone puffs a cigarette might offer insight into the person's inner thoughts (Sonnenschein, 2001).

Yet, the risk with incongruence lies in veering into implausibility or even ridiculousness. Therefore, it may be best to use it sparingly unless the entire experience is intentionally designed around mismatches from the outset, allowing participants to adjust to that framework. A well-crafted narrative might oscillate and strike a balance between congruence and incongruence—challenging the listener, offering surprises, and inviting them to play along and suspend their disbelief.

In *The Reign Union*, intentional incongruence is employed in several scenes. In one (purely fictional) scene, the other main character, Sylvia, tells a story about their dog, 'Puppe'. In the story, her husband, Mauno, had begun using the same name to refer to Sylvia herself. To highlight how demeaning this must have felt, the participant hears the other participant augmented with dog-like sounds—panting and the patter of paws on the floor. This dissonance is designed to spark deeper reflection on Sylvia and Mauno's relationship, exploring themes of patriarchy, sexism, and broader societal dynamics.

Acousmatic transformations and translocations—when you stand in a gallery space but hear a garden around you—are also incongruences. Room divergence, the mismatch between real and virtual acoustics, can degrade externalisation and spatialisation in binaural listening. However, this may be an acceptable trade-off for a meaningful narrative idea. The content creator must rely on the participant's imagination, encouraging them to willingly suspend disbelief and engage with the experience. For instance, in Cilia Eren's audio walks, participants are invited to 'see' and 'smell' the imagined environment, enhancing their immersion (Steijn, 2023). Further, if we contextualise the transformation as a dream or a memory, the two overlapping realities are perhaps easier to justify.

9.6 Addition

Addition is an advanced variation of using congruent sounds to fuse a new meaning with *added value*. While in the previous examples, the object may have been silent, in this technique, the real object natively produces sound. A virtual sound is then added on top of it, aiming at the best possible congruence possible. The result should alter the perception of the object, reshaping its narrative, much like in *Nature Soundscapes*, where animal sounds were overlaid onto the forest soundscape, transforming the experience.

Addition requires an acoustically transparent system for the user to be able to hear the object's native sound. With an acoustically isolated system, the object's own sound should be reproduced, hence missing the core idea.

When the object behaves predictably or can be controlled, addition is easy to integrate into the story. With unpredictable and uncontrolled

Table 9.2 Examples of new meanings created through the merging of native and added sounds

Object	Native sound	Added sound(s)	New meaning
Lift	Whir, hum, chime	Creaking cables, grinding noise	Unsafe, outdated, tension, claustrophobia
Wind chimes	Tinkling, chiming	Resonance, harmonic overtones	Mysticism, emotional depth
Closing door	Slam	Scream	Narrative layer

objects, the randomness must be taken into account. For example, in *Walkie-Talkie Dream Angles* at Washington Square Park, NYC, sounds of demonstrations were added to a corner of the square where such events frequently occur in real life (Naphtali and Rodkin, 2019). Even though the system was using head-locked stereo and thus not reaching perfect congruence, in case a real demonstration happened to take place when the participant was listening to the audio walk, a layer of history was juxtaposed on the current-day event, thus potentially adding context to the present-day politics and events.

However, this technique seems challenging to implement, as examples of its use are rare, particularly with individual objects. This may be because objects that produce regular or constant sounds are uncommon in many exhibition spaces, or they lack narrative interest. Additionally, as noted earlier, irregular and unpredictable sounds are inherently difficult to incorporate effectively.

9.7 Alternative match

Alternative match involves augmenting a real-world object with multiple alternative sounds, each fully congruent with the object but conveying entirely different meanings. This approach can be used to provide different users with alternative versions of a story, such as presenting events from varying characters' perspectives. For example, as a bad smell begins to fill the room, one user hears the sound of a car engine idling outside, implying the smell comes from the tailpipe, while another hears a hiss, suggesting a gas pipe might be leaking. The users' reactions would likely differ significantly based on the sound and meaning they are presented with.

These alternative matches would be similar to the seminal example from Chris Marker's 1957 film *Letter from Siberia*, where documentary images of Yakutsk, Russia, are offered completely different meanings through alternative narrations. Similarly, a single participant in an AAR experience could hear the same event from multiple narrative angles. The system could select the perspective based on the user's behavioural patterns, but these angles could also be changed by moving across the space or around an object, as if the perspective was literally changing. In a study by Cliffe et al. (2021), a physical radio receiver emitted four separate virtual broadcasts simultaneously, each hearable from different sides of the radio. This embodied interaction encouraged participants to explore and discover the unexpected.

Alternative match is related to the concept of *asymmetric information* that was studied, for instance, in *Please Confirm You Are Not a Robot*.

There, participants heard different elements of the story, which consequently made them give accounts to each other of what they heard, prompting group coordination and discovery (Nagele et al., 2021).

The use of alternative match is potentially very powerful in narrative AAR since the surrounding environment is inherently 'real', seamlessly fulfilling its role in the illusion. The challenge, of course, lies in creating virtual audio content—all its alternative versions—congruent enough to maintain plausibility. Beyond art and entertainment, applying this concept to ubiquitous AAR raises intriguing and even unsettling possibilities: could the world sound different depending on an individual's personal interests, political views, income level?

9.8 Extension

Extension is Michel Chion's (1994) term referring to the space beyond the visual field created and suggested by sound. The amount of extension can vary. When the extension is *null*, no sound can be heard around the event we focus on. In *vast extension*, we are able to hear not just the room around us, but the whole building and city.

This concept can be considered central to augmented reality, and in AAR, it is a powerful narrative technique. Using primarily acousmatic sounds, extension can be used to, at least, 1) suggest location, environment, time, weather, and other external parameters outside of the visual or otherwise perceivable space, and 2) set the narrative focus. The natural surroundings behind obstacles such as walls, trees, and buildings, can be extended through sound, or if the technology allows, be acoustically attenuated and even removed.

As discussed in Chapter 4.2, *The Worlds' Deepest Bin* advertisement campaign equipped a public litter bin with a motion sensor and concealed speakers. When a passer-by threw trash inside the bin, a long descending whistle was heard as if the bin extended hundreds of metres underground. As an example of a more realistic and likely more plausible extension, in *The Reign Union*, the participant hears sounds through the walls as if coming from adjacent apartments—laughter, a gramophone playing, someone practicing piano—even though, in reality, the premises have no immediate neighbours.

In the previous examples, imaginary spaces and locations are created beyond the observable reality. At *Maison Gainsbourg*, the visitor starts to hear rain patter hit the roof as if it had started to rain, creating a comfortable and safe atmosphere, suggestively reflecting how the children had experienced their decadent home. If the user is unaware of what is behind the perceivable area—or outside of their field of view—

the surroundings can be turned into anything through sounds, like the *Zombies, Run!* adding distant breathing of zombies behind the user.

If the AAR system allows sound removal, null extension becomes possible. With experiences using acoustically isolated headphones, the attenuation of surrounding sounds happens naturally. This may potentially help the user to focus on the content at hand, especially in noisy environments. The narrative possibilities are obviously significant, too, especially if the ambient sounds can be dynamically controlled. The effect would be similar to the infamous Omaha Beach scene from Steven Spielberg's 1998 film *Saving Private Ryan* where almost all sounds are muted, thus focusing on the protagonist's internal experience.

9.9 Revelation

In *revelation*, the source of an obstructed sound gets revealed. This is an old sound design technique used in at least theatre, opera, and cinema when a character is first introduced through an off-stage or off-screen voice, and is shown only later. The technique is potentially very effective also in AAR. It has a relation to Chion's (1994) concept of *deacousmatisation*, a process where a somewhat mystified and faceless off-screen sound gets embodied with its identity being revealed.

The concept works best if the participant is offered a surprise, revelation, joke, or a counterpoint. This happens onboard *Barque Sigyn* when the source of the cat meow and sound of a dropping object are revealed in the next room. Another example is from *The Reign Union*, where the participants hear the two virtual characters from another room as *obstructed* objects, without seeing them. While Sylvia stays acousmatic mostly throughout the story, Mauno gets revealed: he calls the participant to come to the backroom, and after they have followed his voice, the source of his voice is finally revealed as a photograph in a picture frame. Mauno, who until this point seemed mysterious and somewhat powerful, is truly demythologised, imprisoned inside the picture frame.

A third example uses direct augmentation and is similar to the cat meow onboard *Barque Sigyn*. At one Christmas when our daughter was four years old, Santa Claus was too busy to personally visit our home, but had sent his elves to bring the presents: when we were sitting at table eating the dinner, there was a knock on the door. After a moment we heard another knock. Our daughter's eyes widened, and she went for the door. When opening the door, there was no-one there, but instead the 'elves' had left us a burlap sack filled with presents and hurried off for the next home.

The technology behind the deception was simple. I had hidden a Bluetooth speaker close to the door. During the dinner, I secretly opened my cellphone that was connected to the speaker and triggered a knock sound taken from a sound effects library. The trick apparently worked with the revelation part being an essential component in creating the illusion.

9.10 Use of acousmatic sounds

As suggested earlier, in the context of AAR, acousmatic sounds refer to sounds without a physical counterpart. An acousmatic sound requires using imagination to come up with an idea what the sound source is and what it could look, or smell, or feel like (Chion, 1994; Smalley, 1997). In the *Guided by Voices* game project, the players were physically moving through a real-world space while interacting with acousmatic characters and creatures, such as a dragon. The game took the advantage of the players' imagination, and as the authors state, 'instead of being presented with a rendered dragon that may or may not be frightening, the player gets to interact with his/her idea of what a frightening dragon should look like' (Lyons, Gandy and Starner, 2000).

In *The Reign Union*, Sylvia presents herself to the participant as an acousmatic character. In her dialogue, or monologue to be more precise, she asks the participant to take a look at her clothes and appearance until she remembers and acknowledges her invisibility. The dialogue attempts to create an interactive relationship with the participant, and encourages the use of imagination to enhance the sense of (Sylvia's) presence.

One might think that the use of acousmatic sounds takes us away from audio *augmented* reality and moves us towards audio *virtual* reality. While conceptually that may be true in some cases, acousmatic sounds can still be tightly within the AAR realm at least for two reasons:

1   Even though an acousmatic sound does not have a physical appearance, it can still coexist with the real environment. An example would be a virtual auditory sticker posted at a certain location in the *real* 3D space (Lokki et al., 2004) or a ghost drifting in the same room with the user, both examples being spatially rooted in the real environment.

2   Due to technical limitations, many AAR systems do not allow accurate spatialisation of virtual sounds, if any spatialisation at all, so

acousmatic sounds may be the only way to augment the environment. While they are not attached to any physical objects, they can still appear immersive, especially if supported by strong narrative.

9.10.1 Plausibility

The concept of sound emanating from a point in space without a perceivable source is unnatural. As Chion (1994) points out, we seldom ask *where* the sound is but where it is coming from. We often hear sounds that we cannot see: a distant rumble of a truck, a family member in another room, telephone ringing under a pile of clothes. However, it is rare to hear a sound in mid-air within our reach, a sound that we can circle around but still not see, touch, or smell. Perhaps, a mosquito flying around the head, always avoiding one's hand, would qualify as an example, but eventually we would get a glimpse of it or feel it either between our palms or as an itching sting after a while.

Acousmatic sounds may be hard for the user to understand and localise, unless their source is first seen—or otherwise perceived—after which they become acousmatic (Armbruster, 2024). That technique is called 'detachment', to be discussed in the next subchapter. On the other hand, it may be enough to first introduce the character as obstructed, behind a door or talking from another room, and after that, to reveal them as being immaterial.

Another way to potentially improve the plausibility of an acousmatic character is to let them acknowledge the presence of the participant and invite the participant for interaction, *Hyperkuulo* being a good example. Believable acting may also be important, leading to an overall believability. When recording actors, however, it may be difficult to create an optimal setting for authentic acting. As the object-based audio (OBA) approach requires dry recordings, it may be hard to arrange the actors room to move around and physically act out the scene. If the audio system can tolerate—or even benefit from—leakage of room acoustics into the recording, actors are likely to deliver a more authentic performance. This is a technique leveraged by DARKFIELD in their productions, where they use static dummy head recordings in real acoustic environments to enhance authenticity (Kadel and Rees, 2024).

Moreover, the use of Foley, such as footsteps and clothe sounds, probably enhances the plausibility of acousmatic sounds. Additionally, utilising the directionality of the sound emitter can enhance realism. For instance, in the case of a virtual character, the sound's tone would change depending on the direction they are facing. The character's movements can be pre-animated using various techniques or controlled in real-time by an AI-based system that moves and turns the virtual character dynamically. Finally, high-quality virtual audio rendering can further enhance the overall realism and immersion.

9.10.2 Unavoidable but powerful

Acousmatic sounds seem to be common in narrative experiences. *Guided by Voices, Audio Legends, Séance, Sonic Traces Heldenplatz, Maison Gainsbourg*, and *Corvette Karjala* are just some examples from this book where acousmatic characters and other sounds convey the story. At *Maison Gainsbourg*, the use of acousmatic sounds is very powerful as the visitor hears real recordings of the people who have lived in the house, and even recorded in the same rooms where they are played back after several decades. For instance, the visitor hears a conversation between Serge and his children in the kitchen, or how Serge teaches piano to his daughter Charlotte in their living room, spatialised at the grand piano. Not many other media are capable of so powerfully transporting past moments into present time.

The use of acousmatic sounds may first seem as a 'lazy' way to do AAR, since no effort must be made to find physical, real-world objects that could be augmented with narrative sonic content. Then again, complex narratives may be impossible to realise without relying on acousmatic techniques. And, as *Maison Gainsbourg* demonstrates, these auditory 'ghosts' can sometimes hold enormous narrative powers.

9.10.3 Reference points

It is not a trivial task to create acousmatic virtual sounds that appear as originating from a clearly perceivable point in the environment, especially in 6DoF applications. For binaural systems, high-quality tracking and virtual audio engine are required. Wave field synthesis (WFS) is capable of creating accurately localised 'holophonic' sounds, and experiments have been also done using two directional speakers utilising air nonlinearity (Zhou et al., 2024). When such an illusion has been created, it feels magical and real at the same time. That said, acousmatic sounds may be more forgiving in terms of their positional accuracy compared to sounds that are supposed to be attached to an object—acousmatic sounds are, in the end, 'ghosts' than can slightly float around.

In cases where the application is available to spatialise sounds within the environment, one must remember that acousmatic sounds are often anchored to some object, even if it is an invisible one. For instance, at *Maison Gainsbourg*, a recording of Serge giving a piano lesson was attached around the piano bench, and the conversation between Serge and his children was aligned in the middle of the kitchen. Thus, there is often a reference point in the environment. If unclear, this reference point can be pointed out in dialogue. In the beginning of *The Reign Union*, the acousmatic character of Sylvia enters the room and starts to walk through the space. Even though the participant can follow her spatialised voice and footsteps, she keeps marking her position in the dialogue while moving: 'Maria, *these laundry bags* need to be moved away from *the stairs*. ... *This copper lamp shade* should be replaced, too. It's a bit too modern for this place. ... It's quite draughty here. I should replace the weather stripping on *the vestibule door*.' In another scene, when Sylvia enters a small side room, we hear her turn on a light switch, and at the same moment, a software-controllable real-world light turns on in the room. That is another example of underlining where the acousmatic character is supposed to be.

9.10.4 Acousmatic sounds through loudspeakers

The concept of acousmatic sounds is not solely the property of binaural, WFS, or sophisticated directional speaker techniques. One can achieve the effect of invisible, 'ghost-like' sound sources by using normal hidden loudspeakers. For instance, on the bridge of *Corvette Karjala*, the visitor can hear a conversation between the officer of the desk and helmsman.

The officer's voice is 'acousmatic', appearing around the navigation table without a very precise location. The small loudspeaker is concealed under the radio navigation device (see arrow in Figure 9.3). The sound is, of course, not a point source floating in the mid-air, but still, as the small loudspeaker spreads the sound around and it gets diffused by nearby surfaces, its precise origin gets blurred. Consequently, it is surprisingly easy to imagine the officer standing in front of the table.

The helmsman's voice is, however, attached to the speaking tube above the rudder post: the loudspeaker is placed inside the tube. The other end of the speaking tube is on the sky bridge above; perhaps the narrative suggests that the helmsman is steering the vessel from there. Even with such a simple realisation using two loudspeakers and a looping, two-channel audio file, the dialogue effectively immerses the visitor in the life on the bridge and educates about the communication practices of a naval vessel.

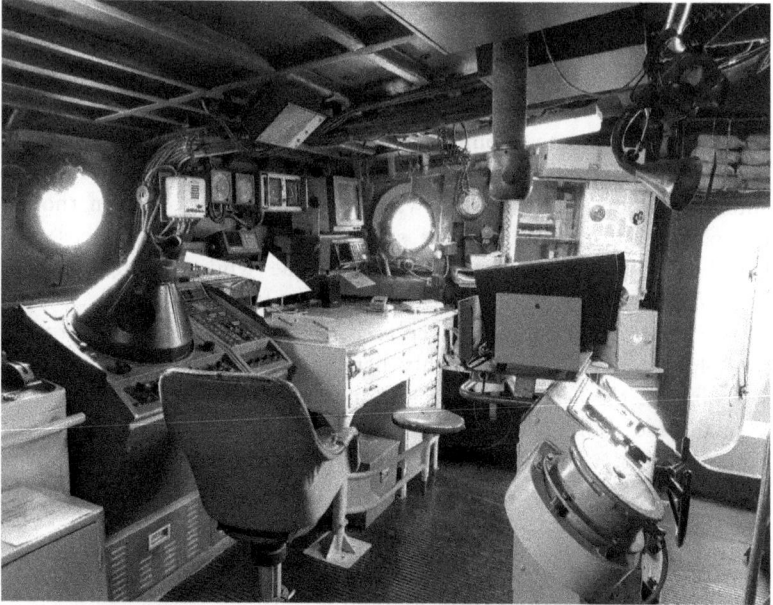

Figure 9.3 The bridge of *Corvette Karjala*. Arrow pointing to the hidden loudspeaker
Source: Photograph by the author

9.11 Detachment

In this technique, the virtual sound is first attached to a physical object, after which it gets *detached*. At that moment, it becomes acousmatic, 'incorporeal' sound. This can mark a change of narrative perspective, for instance, a move from the tangible world to the inner world of a character. In *The Reign Union*, Mauno, who is locked inside a picture frame, starts to play a gramophone record, the sound of which is also attached to the frame. In the middle of the song, Mauno asks the participant to join the dance by grabbing the picture frame. When the picture frame is lifted, the music gets spatialised around the room as individual, acousmatic instruments. The effect is spectacular, as if we were sucked inside Mauno's head, surrounded by his carnivalesque imagination.

Detachment is similar to Chion's (2019) concept of *acousmatization*: 'with the effect that we now have nothing but the sound to imagine what is occuring' (p. 201). It is the opposite of 'deacousmatisation', discussed earlier in the context of the revelation technique.

9.12 Attractor sounds

The key technique to directing the user's attention across the space is to use *attractor sounds* as explored in *LISTEN* (Zimmermann and Lorenz, 2008). An attractor sound would be a virtual sound emitting from an object or place, distinguishable from the other sounds and intriguing enough to make the participant take an action to find out what the sound will reveal. Attractor sounds can also be triggered adaptively and personalised to each user as demonstrated in *LISTEN*. Due to the physical attributes of sound, it would be entirely plausible to hear attraction sounds from obstructed objects, other rooms, or behind obstacles, encouraging exploration beyond the immediate environment.

When using an attractor sound, two trigger zones can be created around the source object, sometimes called *proximity zone* and *activation zone* (Vazquez-Alvarez et al., 2016). Proximity zone is larger, activation zone smaller. When the participant enters the proximity zone, or is already inside it, the attractor sound is played—for example, a telephone ringing behind a door. If the participant keeps stationed for a set period of time or moves out of the zone, the sound may stop to allow further exploration or alternative progression of the narrative. However, if the participant decides to move closer to the sound, the activation zone gets triggered, and the actual content starts to play. The transition must, of course, be smooth and natural—unless the attractor sound can continue playing in the background with the content mixed on top of it. Behind the door, we may hear footsteps approach, someone pick up the telephone, and a conversation start.

Instead of using map-based trigger zones, sounds can be controlled based on the user's distance to the object. This approach is an obvious choice when only proximity tracking is available but can also work effectively with absolute tracking systems. For instance, distance-based triggering is a built-in feature in most audio middleware.

Attractor sounds are useful in exploration-based scenes where the participant can hear sounds from around the surroundings—invitations to narratives—and decide which one to choose. However, in practice, such a scene may be challenging to balance and programme for at least two reasons:

1   The attractor sounds should be audible far enough for the participant to hear at least two of them simultaneously. If only one is heard, the user will likely go directly to that source, skipping the exploration phase. However, if multiple sounds are audible, there is a risk of cacophony, so attenuation curves must be adjusted carefully and the sounds selected and mixed with care.

2   The logic of handling the different attractor sounds and the narrative content found from their sources may get complicated. For instance, what if the user selects one attractor and starts to listen to the content, but has second thoughts and walks away? Should the narrative continue and be audible to the user, fade out and pause—or fade out, stop and rewind? Meanwhile, what happened to the other attractor sounds; did they disappear, or are they still active? Furthermore, if the scene has a time limit after which the next one begins, users who explore multiple sounds may not have enough time to finish listening to their chosen story. These are, of course, common challenges in developing interactive experiences and these are not unique to AAR.

9.13 Spatial scale and offset

Adjusting attenuation curves is a common technique in sound design, typically used to define the distance over which sounds can be heard. This, in a way, manipulates the spatial scale of the auditory perception of the individual sounds from a 'natural' state to an artificial one. There are also some other ways to adjust the auditory perspective and scale with potential narrative benefits.

9.13.1 Manipulated attenuation curves

The attenuation curves can be adjusted, for example, to prevent simultaneous sounds from overlapping, ensure that certain sounds are only audible from closer distances, or allow some sounds to remain constant regardless of the user's distance from them. Even though manipulated attenuation curves make virtual sounds behave contrary to the laws of physics, the result may provide a more coherent and controlled soundscape that, within its context, can feel natural and plausible to the listener (Cliffe, 2022).

The curves can also be manipulated for more creative purposes. The concept of 'zooming' into a distant sound source is a prime example. In *Hyperkuulo*, for instance, participants at one point could hear the sounds of a distant public sauna as if they were much closer. While this effect can be achieved through various technical methods, radically adjusting attenuation curves offers a simple and effective solution.

Martínez-Cabrejas et al. (2024) propose a *directional attenuation model* where only the sounds that the user is facing will be heard while the surrounding sounds get attenuated. The idea is not new, and while it could help users focus on objects of interest, extreme care should be practiced when realising such a system: as Armbruster (2024) points out, in exhibitions, people keep turning their head and body erratically— perhaps looking for their friend and asking them to listen—and filtering out such user movements would require sophisticated algorithms.

9.13.2 Scaled spatial response

In normal 3DoF, 6DoF, and XY-Yaw systems, turning the head 90 degrees to the right would counter-rotate the virtual soundscape 90 degrees to the left, thus keeping the sounds aligned with the real world. However, for some sounds, it may be justified to scale down the spatial response: in a 90 degree head turn, the sound would pan only, say, 45 degrees (see Figure 9.4). The *Vienna Augarten* experience, where proximity to porcelain exhibits trigger voice descriptions, utilises this technique of *scaled spatial response*: the narrator's voice is not fully spatialised to the exhibits, nor is it fully head-locked either, but it is moderately panning towards the target object. This maintains cognitive focus on the description while still connecting the narration spatially to the exhibit. This technique could work for other non-diegetic sounds, too, such as music.

9.13.3 3DoF sounds

Sounds embedded in the environment are usually *allocentric*, that is, relative to the environment, and head-locked sounds are, in turn, *egocentric*, following the listener's movements and head orientation (Tsingos, 2018). However, it can sometimes be beneficial to have sounds

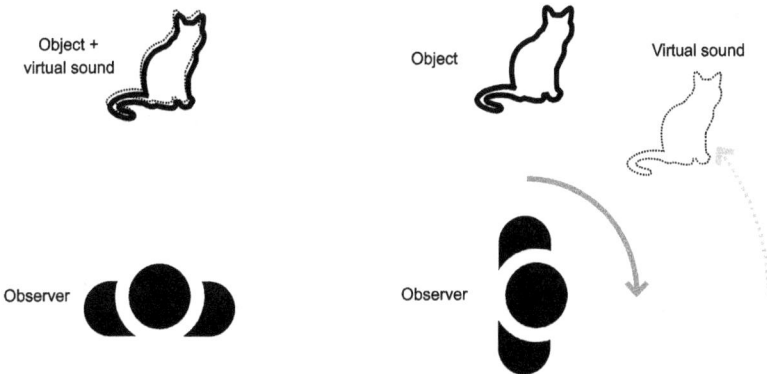

Figure 9.4 Scaled spatial response

move with the user while still maintaining head tracking, thus utilising three degrees of freedom (3DoF). Ambisonic ambiences sometimes use this 3DoF attachment, and the technique can also be applied to object-based audio sources, for example, by creating a dome of virtual speakers around the user. This approach is justified in mobile apps and other applications that allow free movement across an undefined place. Also, narrative and sound design reasons may call for it, for example, in memory scenes where the real environment is contextually 'faded' to the background.

9.13.4 Spatial offset

The spatial coordinates of virtual sounds can also be relative to entities other than the real environment or the user. In real life, the sound of an airplane in the sky often seems to lag behind its visual position. This phenomenon, caused by the slower speed of sound compared to light, can create the illusion of an offset auditory world. Conversely, the sound of an approaching train may appear to emanate from the rails ahead of the train—with its distinct laser gun-like effect. These kinds of effects used in AAR can work narratively to suggest we are perceiving the world from a different auditory perspective—either from our own standpoint but with an offset, or fully aligned with another frame of reference. For example, in the theatre play *Anna* at the National Theatre in the UK, the audience, wearing headphones, experienced stage events captured by binaural microphones placed in the protagonist's ears. While visually observing the events from their seats, they heard the auditory world as if it were attached to the actor's head. A similar spatial offset can occur in binaural audio walks if the participant deviates from the intended tempo or path.

The airplane and train examples could also imply a temporal offset, where the listener hears the world events either before or after they happen in other modalities. While these offsets offer intriguing possibilities for AAR, such effects naturally risk being perceived as errors if they are not well-justified within the narrative context.

9.14 Near field

Near field sounds refer to sounds that are very close to the listener's head, typically closer than one metre, but still externalised, not playing inside the head. For many people, they trigger a physical reaction that is often considered pleasing or otherwise strong, as is shown by the popularity of the ASMR videos. With a human voice, the feeling may be very intimate, or depending on the context and voice, intimidating.

In AAR, near field sounds can be realised at least with binaural systems. When positioning the sounds out of the user's field of view—on sides, behind, above and under the head—it is possible to create a plausible sensation of them really being there. Even head-tracking is not necessarily needed, especially if the sound is constantly moving, like a mosquito, or it is supposed to move with the listener's head, like a spirit whispering into their ear.

In *Séance*, near field sounds are used powerfully. They can provide a perspective shift from the environment to the 'invironment', the intimate and immediate private space of ourselves (Mann et al., 2023). This effect gets enhanced in audio MR, or AAR experiences where virtual sounds are almost indistinguishable from the reality—we cannot be completely sure if the sound right behind our ear is real or not.

9.15 Removal and replacement

While selectively removing sounds is not yet a tool readily available for AAR creators, it will definitely offer interesting narrative possibilities. As discussed, 'acoustic intrusions' or disturbing sounds are a challenge, which could be eliminated with adaptive and selective noise cancelling. Such technology is sometimes called *diminished audio* (Dam et al., 2024), and it could also find more creative uses.

The deliberate omission of sound for an object could be effective by drawing attention to its silence as a narrative or emotional tool. For example, a music box could fall silent when the listener approaches it, perhaps symbolising vulnerability or solitude.

The null extension technique discussed earlier could also benefit from audio removing, especially when only selected sounds were to be moved from the environment. Further, a person's voice could be erased and replaced with scripted, virtual speech. This opens both creative and frightening possibilities, a door to digital deception already witnessed with deepfake videos (Arena et al., 2022; Turner, 2022). In the 'shred' videos on the Internet, the original audio of a live concert video by a world-famous artist is replaced with some-one's hilarious re-interpretation recorded in their bedroom using cheap MIDI instruments. Whenever a musician is out of frame, their instrument stops playing, making the music sound fragmented and amateurish. The live performance context enhances the illusion of 'real', and adapting the same concept to AAR would be extremely intriguing.

9.16 Synchronisation

Since AAR merges virtual and real, synchronisation between these two realms is essential. Every interaction in AAR is, in a way, an instance of synchronisation. However, here the concept refers to short, sparse moments where the virtual and physical worlds intersect in a particularly meaningful way. Narratively, there are many interesting approaches to synchronisation, two of which will be briefly discussed here.

The first approach is closely related to Chion's (1994) concept of *synchresis*, where a mental image is created as sound and visuals occur simultaneously, typically within a brief moment. In AAR, this could be achieved by letting the real world control virtual audio events, precisely synchronising a virtual sound to a single, momentary real-world event in order to create a contextual fusion between the two. For example, when the dials of a wall clock strike 12 o'clock, the bustling street sounds heard through a window could suddenly shift to serene nature sounds, perhaps symbolising a moment of release or freedom.

Synchronisation would potentially be most effective when the real-world event is unpredictable and caused by a real object, such as other people, animals, nature, etc. However, implementing such synchronisation may be difficult to design and would require an advanced sensory system.

Alternatively, the user can be guided to act in synchronisation with virtual audio. In Rotozaza's *Etiquette,* a participatory theatre performance staged in real cafés, two participants wear headphones and receive instructions on what to do. At one point, one participant is prompted to place their fingers on the table at the cue 'One, two, three, now'. At 'now', as the fingers presumably touch the table, a piano chord plays (Rotozaza, 2015). While the success of this effect relies on the participant's cooperation, it mirrors the concept of synchronising actions with an auditory protagonist, as demonstrated in Janet Cardiff's audio walks and the *Alcatraz Cellhouse Tour,* creating a direct and tangible link between the virtual and physical world.

9.17 Acoustic translocation

Acoustic translocation is a technique where the real space is acoustically transformed to a virtual one by introducing new ambience sounds and auditory story elements. An office space can take on the sounds of a railway station, with station announcements echoing and platform whistles signalling departures. Being able to change the location of events would obviously be beneficial to many narratives. Yet, that would be against the nature of augmented reality: if we 'deny' the real space, we take a step towards the virtual reality domain.

That said, even if we introduce a translocated soundscape, we cannot get rid of the real environment. There will likely be a mismatch between what the participant *hears* coming from the virtual world and how the real environment advertises itself through *all* of the senses, including hearing—unless an acoustically isolated or adjustable system is used. If there is a conflict between image and sound, the visual modality tends to win (Stevens, 2009), backed up by smell and other senses of the real space. Therefore, acoustic translocation primarily functions on a contextual level, offering a more or less symbolic representation of another place. While the listener can—and likely often will—imagine seeing and smelling the translocated space, we can assume that these perceptions remain highly subjective.

However, when acoustic translocation is viewed as a method of *reframing* the context of the environment (see Cliffe et al., 2021) rather than an attempt to fully transport the participant to a new space, it aligns more naturally with the principles of AAR. For instance, overlaying the soundscape of a railway station onto an office environment could symbolically highlight workplace transitions if not fusing deeper meanings.

Technically, the room divergence poses a challenge as the mismatch between real and virtual acoustics may cause externalisation collapse and spatialisation to get blurry. The railway station with a longer reverberation time than the office may not pose a significant problem (e.g., Brandenburg and Sloma, 2024), but overlaying an outdoor scene with a shorter reverberation time could, for example, make distance perception challenging. One way to better 'glue' virtual sounds to the real environment is to incorporate some amount of real-space acoustics into the virtual sounds, thereby smoothing out the psychoacoustic mismatch. Practical experience suggests that this approach works to some extent, though its effectiveness is undoubtedly context-dependent, highlighting the need for further research in this area.

In some AAR experiences, however, the virtual scene keeps evolving independently, and the real world becomes a mere physical platform on which the auditory narrative takes place (e.g., *Guided by Voices* and *The Planets*). We can compare this to music videos, where the visually conveyed narrative is often open, fragmented, or even surrealistic in order to keep the listener's focus on the music while allowing the imagination to create individual meanings (Stevens, 2009). In music videos, there are two narrative layers, music and images, which have their own freedom. In AAR, to keep the events in the virtual 'elsewhere' rooted in the real environment, *points of synchronisation* could be arranged every now and then, just like in music videos where the band is seen to perform the song between other imagery (Chion, 2019). In AAR, the virtual characters in a translocated scene could react to events happening in the real environment, like protagonists in *The Matrix* films. There can obviously be many other 'flows' between these two narrative layers, like objects of the real world becoming objects of the virtual place with the same or slightly changed meaning.

9.18 Conclusion

Observing the emergence and evolution of a new medium is an exciting process. With rapid technological advancements and increasing accessibility, almost anyone can take part in this process and contribute to shaping and advancing AAR. The narrative techniques and concepts presented in this chapter aim to inspire readers to engage with and interpret AAR content from their own perspectives. Additionally, it is sincerely hoped that these ideas will spark conversations and serve as inspiration for practitioners as they continue to develop the narrative language of AAR.

References

Albrecht, R., Väänänen, R. and Lokki, T. (2016) 'Guided by Music: Pedestrian and Cyclist Navigation with Route and Beacon Guidance', *Personal and Ubiquitous Computing*, 20 (1), pp. 121–145. Available at: https://doi.org/10.1007/s00779-016-0906-z.

Arena, F. et al. (2022) 'An Overview of Augmented Reality', *Computers*, 11 (2), p. 28. Available at: https://doi.org/10.3390/computers11020028.

Armbruster, S. (2024) 'Conversation with Matias Harju, March 7'.

Brandenburg, K. and Sloma, U. (2024) 'Conversation with Matias Harju, May 23'.

Chion, M. (1994) *Audio-Vision: Sound on Screen*. Columbia University Press.

Chion, M. (2019) *Audio-vision: Sound on Screen*, 2nd edn. Columbia University Press.

Cliffe, L. (2022) *Audio Augmented Objects and the Audio Augmented Reality Experience*. PhD thesis. University of Nottingham. Available at: https://eprints.nottingham.ac.uk/69795/.

Cliffe, L. (2024) 'Into the Here and Now: Explorations within a New Acoustic Virtual Reality', *Leonardo* [Preprint]. Available at: https://nottingham-repository.worktribe.com/output/42480431/into-the-here-and-now-explorations-within-a-new-acoustic-virtual-reality (Accessed: 2 December 2024).

Cliffe, L. et al. (2021) 'Materialising Contexts: Virtual Soundscapes for Real-World Exploration', *Personal and Ubiquitous Computing*, 25 (4), pp. 623–636. Available at: https://doi.org/10.1007/s00779-020-01405-3.

Cohen, M., Aoki, S. and Koizumi, N. (1993) 'Augmented Audio Reality: Telepresence/VR Hybrid Acoustic Environments', in *Proceedings of 1993 2nd IEEE International Workshop on Robot and Human Communication*, pp. 361–364. Available at: https://doi.org/10.1109/ROMAN.1993.367692.

Dam, A. et al. (2024) 'Audio Augmented Reality Using Sonification to Enhance Visual Art Experiences: Lessons Learned', *International Journal of Human-Computer Studies*, 191, p. 103329. Available at: https://doi.org/10.1016/j.ijhcs.2024.103329.

Fry, G. (2019) *Sound Design for the Stage*. The Crowfood Press.

Härmä, A. et al. (2004) 'Augmented Reality Audio for Mobile and Wearable Appliances', *Journal of the Audio Engineering Society*, 52 (6), p. 22.

Kadel, O. (Host, Producer) and Rees, E. (Producer) (2024) 'Episode 104 David Rosenberg & Glen Neath (DARKFIELD)'. (Immersive Audio Podcast). Available at: https://immersiveaudiopodcast.com/episode-104-david-rosenberg-glen-neath-darkfield/ (Accessed: 11 October 2024).

Kaghat, F.-Z. and Cubaud, P. (2010) 'Fluid Interaction in Audio-Guided Museum Visit: Authoring Tool and Visitor Device'. The Eurographics Association. Available at: https://doi.org/10.2312/vast/vast10/163-170.

Kuhn, A. (1985) 'History of Narrative Codes', in P. Cook (ed.) *The Cinema Book*. London: BFI, pp. 207–220.

Lawton, M., Cunningham, S. and Convery, I. (2020) 'Nature Soundscapes: An Audio Augmented Reality Experience', in *Proceedings of the 15th International Audio Mostly Conference (AM'20)*, Graz Austria: ACM, pp. 85–92. Available at: https://doi.org/10.1145/3411109.3411142.

Lokki, T. et al. (2004) 'Application Scenarios of Wearable and Mobile Augmented Reality Audio', in *Audio Engineering Society Convention* 116, Audio Engineering Society, p. 9.

Lyons, K., Gandy, M. and Starner, T. (2000) 'Guided by Voices: An Audio Augmented Reality System', in *Proceedings of the International Conference on Auditory Display April, 2000*. Georgia Institute of Technology, Atlanta, Georgia, USA: International Community for Auditory Display. Available at: https://smartech.gatech.edu/handle/1853/50672 (Accessed: 5 November 2018).

Mann, S. et al. (2023) 'Fundamentals of All the Realities: Virtual, Augmented, Mediated, Multimediated, and Beyond', in A.Y.C. Nee and S.K. Ong (eds) *Springer Handbook of Augmented Reality*. Cham, Switzerland: Springer (Springer Handbooks).

Martínez-Cabrejas, M. et al. (2024) 'A New Listener-Centered Directional Attenuation Sound Model for Augmented Reality Environments', *Multimedia Tools and Applications* [Preprint]. Available at: https://doi.org/10.1007/s11042-023-17943-w.

Nagele, A.N. et al. (2021) 'Interactive Audio Augmented Reality in Participatory Performance', *Frontiers in Virtual Reality*, 1, p. 610320. Available at: https://doi.org/10.3389/frvir.2020.610320.

Naphtali, D. and Rodkin, R. (2019) 'Audio Augmented Reality for Interactive Soundwalks, Sound Art and Music Delivery', in *Foundations in Sound Design for Interactive Media*. Routledge.

Pick, H.L., Warren, D.H. and Hay, J.C. (1969) 'Sensory Conflict in Judgments of Spatial Direction', *Perception & Psychophysics*, 6 (4), pp. 203–205. Available at: https://doi.org/10.3758/BF03207017.

Rotozaza (2015) 'ETIQUETTE reviewed by New York Times'. Available at: https://vimeo.com/120873104 (Accessed: 10 December 2024).

Smalley, D. (1997) 'Spectromorphology: Explaining Sound-Shapes', *Organised Sound*, 2 (2), pp. 107–126. Available at: https://doi.org/10.1017/S1355771897009059.

Sonnenschein, D. (2001) *Sound Design: The Expressive Power of Music, Voice, and Sound Effects in Cinema*. Studio City, CA: Michael Wiese.

Steijn, R. (2023) 'Ode aan de verstilling – Cilia Erens (1946–2023)', *Theaterkrant*. Available at: https://www.theaterkrant.nl/tm-artikel/ode-aan-de-verstilling-cilia-erens-1946-2023/ (Accessed: 9 October 2024).

Stevens, M. (2009) *Music and Image in Concert: Using Images in the Instrumental Music Concert*. Sydney: Music and Media.

Tsepapadakis, M. and Gavalas, D. (2023) 'Are You Talking to Me? An Audio Augmented Reality Conversational Guide for Cultural Heritage', *Pervasive and Mobile Computing*, 92, p. 101797. Available at: https://doi.org/10.1016/j.pmcj.2023.101797.

Tsingos, N. (2018) 'Object-Based Audio', in A. Roginska and P. Geluso (eds) *Immersive Sound: The Art and Science of Binaural and Multi-Channel Audio*. Routledge, pp. 244–275.

Turner, C. (2022) 'Augmented Reality, Augmented Epistemology, and the Real-World Web', *Philosophy & Technology*, 35 (1), p. 19. Available at: https://doi.org/10.1007/s13347-022-00496-5.

Vazquez-Alvarez, Y. et al. (2016) 'Designing Interactions with Multilevel Auditory Displays in Mobile Audio-Augmented Reality', *ACM Transactions on Computer-Human Interaction*, 23 (1), pp. 1–30. Available at: https://doi.org/10.1145/2829944.

Väänänen-Vainio-Mattila, K. et al. (2013) 'User Experience and Usage Scenarios of Audio-Tactile Interaction with Virtual Objects in a Physical Environment', in *Proceedings of the 6th International Conference on Designing Pleasurable Products and Interfaces*. New York, NY, USA: ACM (DPPI '13), pp. 67–76. Available at: https://doi.org/10.1145/2513506.2513514.

Wakkary, R. et al. (2004) 'Interactive Audio Content: An Approach to Audio Content for a Dynamic Museum Experience through Augmented Audio Reality and Adaptive Information Retrieval'. Available at: http://summit.sfu.ca/item/15157 (Accessed: 5 November 2018).

Yang, Z. et al. (2020) 'Ear-AR: Indoor Acoustic Augmented Reality on Earphones', in *Proceedings of the 26th Annual International Conference on Mobile Computing and Networking. MobiCom '20*, London, UK: ACM, pp. 1–14. Available at: https://doi.org/10.1145/3372224.3419213.

Zhou, J. et al. (2024) 'Visar: Projecting Virtual Sound Spots for Acoustic Augmented Reality Using Air Nonlinearity', *Proceedings of the ACM on Interactive, Mobile, Wearable and Ubiquitous Technologies*, 8 (3), pp. 1–147. Available at: https://doi.org/10.1145/3678546.

Zimmermann, A. and Lorenz, A. (2008) 'LISTEN: A User-Adaptive Audio-Augmented Museum Guide', *User Modeling and User-Adapted Interaction*, 18 (5), pp. 389–416. Available at: https://doi.org/10.1007/s11257-008-9049-x.

10 The future of AAR

The future of AAR echoes with hope and uncertainty. People have grown accustomed to consuming audio content while doing other things—driving, commuting, jogging, doing household chores. In an overly visual world, the rise of personal auditory bubbles filled with audiobooks, podcasts, and endless libraries of music suggests a promising direction for the development of AAR. It seems logical that the earbuds, headphones, and glasses people already wear could evolve into intuitive interfaces for a data-driven, interconnected life. Similarly, as technology advances to create natural-feeling visual and auditory holograms, it seems inevitable that such innovations will bring people closer together. At the same time, virtual acoustic layers could enrich our understanding of the world around us and unlock stories otherwise left untold.

Yet, despite how logical these ideas may seem, their future depends on whether people truly value spatialised, environment-aware, interactive sound. Business plans always return to the same critical question: what problem does this solution address? Is there a genuine need to justify significant investments in technology and product development? Or will the progress of AAR rely instead on creative pioneers willing to push boundaries, even when the tools are imperfect and the results far from polished? After all, many transformative technologies began as solutions to needs people did not yet realise they had—but which have since significantly improved the quality of life.

This book began with *one* vision of personal AAR earbuds. While many other visions for AAR exist, they share a common foundation: the technological components needed to make them a reality already exist, albeit at varying levels of readiness. The main technical challenge lies in integrating these components into compact form factors, like tiny earbuds or glasses, while ensuring high performance and long battery life. Currently, these head-worn devices rely on pairing with smartphones for

DOI: 10.4324/9781003627289-10

offloading computing and utilising external sensors, which remains a potential inconvenience. However, given the high prevalence of people's attachment to their phones, this dependency is unlikely to be a major drawback. Another challenge may lie in inventing and creating content that best utilises the unique features of AAR. However, when thinking about the amount of creativity and skill put into the already realised projects presented in this book alone, that challenge should not turn out to be a major problem.

The opening story in Chapter 1 envisioned several features that a mobile, wearable AAR system could offer to assist people in their lives. These ideas are now revisited with references to the current state of development and general conceptions of future directions.

10.1 Platform

The protagonist in our story is wearing a pair of sophisticated earbuds. In the future, virtual soundscapes can perhaps be 'injected' straight into the auditory cortex in the brain—something currently experimented for tinnitus treatment (Leaver, 2024)—or at least through auditory nerve or brainstem stimulation (see Veronese et al., 2023). In the shadow of that sci-fi vision, however, wearable, binaural transducers seem the most likely platform for AAR. To be used throughout the day, the device needs to be comfortable and socially acceptable. In this regard, hearing aid devices and eyewear work as good examples of potential form factors for ubiquitous AAR. Similar to eyewear, medical hearing aids are designed for continuous use, apart from activities like sleeping or swimming (Andre, 2024). The more common wearable devices get, the more socially acceptable they become, eventually making it perfectly natural to wear them even in social and professional interaction situations.

Smart glasses seem a promising platform for AAR. Their larger size compared to in-ear headphones allows for slightly more room to incorporate processors, memory, batteries, sensors, and antennas, making them a versatile option for advanced AAR functionality. Also, the ability to integrate forward-facing cameras, LiDAR, and microphones—or any future sensors—is likely to make inside-out tracking easier compared to ear-mounted sensors.

One challenge with the current smart glass design is, however, their open-ear speaker design that makes attenuating real-world sounds technically very difficult. Before active sound cancelling technology gets implemented for open-ear systems, hybrid solutions may be used where acoustically isolated earphones complement smart glasses.

10.2 Telepresence

The story started with a teleconference where participants were spatialised around the room as auditory holograms. While the shift away from remote work has gained momentum recently, holographic communication could emerge as a compelling parallel solution, enabling natural collaboration within international teams and reducing the need for travel. Immersive audio can evoke a sense of belonging and connection, enabling natural interactions that bridge physical distances (Yang, Barde and Billinghurst, 2022), and hearing others' voices naturally could result in more seamless and effortless conferencing (Bruhn, Multrus and Varga, 2022). VR conferencing applications such as *Meta Horizon Workrooms* and *Arthur* already spatialise participants' voices. Achieving the same in AAR would involve omitting the avatars and other visual elements and auralising the voices as if they coexisted in the same real-world room with the user (see Meyer-Kahlen, 2024). The constant development of faster mobile data networks, such as the successors for the current 5G, will lead to shorter latencies and better sound quality. One current advancement has been the development of the IVAS codec for low-latency spatial audio in teleconferencing situations (see Chapter 7.8).

Besides teleconferencing, similar low-latency and high-quality spatial audio codecs would enable, for example, hybrid music performances: an orchestra or choir could rehearse and perform in a hybrid formation, blending live and remote musicians. Through telepresence, remote participants would be acoustically present and positioned amongst the physical musicians. With near-zero latencies and high-quality spatial audio codecs, all of the musicians could naturally react and contribute to the performance.

The protagonist ends the call using eye movements, detected by sounds captured with in-ear microphones. Researchers (Lovich et al., 2023) have discovered that eye movements send brain signals back to the ears where they are transformed into tiny sounds, and these sounds have been demonstrated to detect eye movements. While this could be used as a hands-free interaction input, it is also hypothesised that this mechanism may affect sound localisation and other aspects of auditory perception with implications for understanding how the brain integrates information from different senses.

As we move towards our future vision, more discoveries like this are likely to emerge, unlocking new technologies and ways to interact with both our devices and the environment.

10.3 Sound spatialisation

To realistically embed virtual sound objects in the real environment, such as the office room in the story, a key factor is the auralisation of the virtual sounds with the same acoustic properties as the real room. Some current devices, like the *Apple Vision Pro* MR headset, scan the geometry of the environment with sensors, predict the acoustic properties of the surface materials, and use this information for geometry-based simulation of the room acoustics (Das, 2023). However, for more realistic results, it may be necessary to acquire the real acoustic response of the room. This can be done by analysing how the existing sounds in the room excite the room and using that information to calculate the room's reverberation characteristics and early reflections (Meyer-Kahlen, 2024; see Chapter 6.4). This blind estimation approach is still in the research stage but, powered with AI technologies, seems a promising path.

A lot of research has been conducted on individualised HRTFs and how they could be acquired from the user. Their role in improving the binaural spatialisation of virtual sound sources is significant, especially in acoustically dry environments (Best et al., 2020). Some devices can already scan users' ears and send the data to a server for calculating the personal HRTFs. However, with 6DoF motion and well-matching room acoustics, the role of these individualised HRTFs is not that crucial (Brandenburg et al., 2023), and thus, skipping the acquisition of them would probably be an acceptable compromise to make the AAR devices and user experiences simpler (Schmalstieg and Hollerer, 2016).

10.4 Pose tracking

Users' head tracking would be essential for any binaural spatialisation of virtual sounds. Already, very accurate and fast tracking can be realised in indoor spaces with pre-installed sensors, and quite accurately even with the inside-out approach as proven by current MR headsets, enabling tracking in arbitrary places. Outdoors, GNSS with the RTK extension also provides accurate tracking, although, in street canyons particularly, the system may become unreliable due to loss of line-of-sight to the satellites and unwanted reflections from buildings. Optical, markerless inside-out systems will likely develop to be more tolerant of changing lighting conditions, and when combined with LiDAR and other sensors, they show potential to become the primary tracking solution.

Many earbuds can already know their orientation with integrated IMU chips. Adding cameras and LiDAR sensors onto the earbuds for location and gesture tracking is not an impossible idea, although it may pose some problems: privacy concerns when cameras are pointing to other persons, and occlusion issues when the user's ears are covered with, say, big hair or a winter cap. The latter can be solved, to some extent, with external sensors installed in smart glasses, a bicycle helmet—as in our story—a vehicle, or some other device. Cameras tend to trigger privacy concerns more than microphones, even though microphones are already capturing the surrounding soundscapes on millions of headphones, earbuds, hearing aids, and smart glasses. If that escalates into a wider societal issue and hinders the development of camera tracking, it may be that it is LiDAR, acoustic tracking, IR cameras, inertial sensors, and some other yet unknown or under-explored approach, that will be the future of pose and gesture tracking in mobile, wearable devices (Yang, Barde and Billinghurst, 2022; Purcher, 2024).

A major practical challenge with wearable devices in general, especially when intended for pervasive use, is that while they may work perfectly in optimal climate conditions, the performance of sensors and batteries may decrease significantly when it is too cold or hot, too humid or wet, or just too dark or bright for optical sensors to work. The ruggedness of wearable medical and military technology demonstrates that technical solutions to at least some of the aforementioned problems already exist, even if they come with an elevated price tag.

10.5 Sound separation and simultaneous interpretation

In our story, the earbuds are capable of isolating individual sounds and manipulating them. Removing the air-conditioner hum would be a simple trick for any present-day headphone-based ANC system, yet a more intelligent system would be needed to recognise and isolate particular voices from background noise, remove reverberation if needed, and keep them spatialised correctly. Some hearing aids and smart headphones are already capable of many of those tasks, aided by AI-driven algorithms and beamforming—that is, using phase and amplitude differences between two or more microphones to focus the sound capture (Signia, 2024; Sonova, 2024; Duong et al., 2017). For example, in a University of Washington prototype, the user looks at another person once, by which the deep-learning system learns their speech traits and is able to isolate that voice from the background sounds (Veluri et al., 2024). Further, the isolated sound does not have to be speech, but any sound the user desires (Veluri et al., 2023).

Once a sound is isolated from the surroundings, it also can be attenuated and removed completely. This personal 'soundscape configuration' could bring improvements to individuals' lives, such as muting the loud snores of the person next to them (Kiss, Mayer and Schwind, 2020), or lead to more troubling 'Black Mirror' visions where elements of reality are erased (McGill et al., 2020), perhaps without the user's knowledge. A more promising vision involves helping individuals with ADHD or sensory sensitivity to maintain focus, for example in school: selective noise cancelling could eliminate distractions, such as giggling friends, while preserving the teacher's voice and other key sounds audible (see Kulawiak, 2021).

Technically, once a sound is erased, it can be replaced with a virtual alternative, which would be no different from other augmentations discussed throughout this book. One application area might be theatres and cinemas where actors' voices could be dubbed into other languages. Hearing aid loops have been, in a way, precursors to this idea—or AAR in general—by feeding enhanced audio content to the users' ears, and personal AAR devices and services could take the idea much further.

Dubbing, of course, traditionally relies on pre-recorded lines performed by actors. Currently, there are, however, several applications for earbuds—coupled with a smartphone—that offer near real-time interpretation (e.g., *Time Kettle W4 Pro* and *Pixel Buds Pro*). They listen to the chosen language, translate it, and use a speech synthesiser to play it to the user. When such features were combined with the aforementioned sound separation and spatialisation technologies, the translated voice could be spatially localised as if coming from the persons' mouths just like in our story.

10.6 Navigation and situational awareness

In navigational tasks, audio guidance has significant benefits as it frees vision and other senses to observe the surroundings, and such services likely develop to utilise spatial sound. As an inspiration to our story, in the prototype by Albrecht, Väänänen and Lokki (2016), the user was guided to the destination by music that was spatialised to waypoints along the route as auditory beacons.

Like navigation, situational awareness will also definitely remain one of the key application domains of AAR in the future. Binaural 3D audio systems in aviation provide significant benefits in terms of, for instance, reaction speed (Kucinski, 2018), and similar AAR systems could help car drivers detect other vehicles in blind spots when, for example, changing lanes (Lylykangas, 2023). Visual interpreters and obstacle avoidance systems for people with visual impairment will likely develop and get integrated into smart glasses or other head-mounted devices, as prefigured by *Envision Glasses* (Envision, 2023). In our future story, the protagonist's personal system detected an approaching bicycle from behind and sonified it with the sound of a bicycle bell. This would require the use of LiDAR or other sensors with 360-degree vision, not unlike the ones used in autonomous cars, perhaps mounted in the helmet as suggested earlier.

While some industrial environments have adopted visual AR (and MR) for various tasks, AAR has largely been neglected so far. Yet, embedded sonified sounds could hold potential in many applications, such as reporting the device status of different parts of a machine in maintenance and diagnostics systems (Yang, Barde and Billinghurst, 2022). Sensory augmentation with spatialised sound could also find uses in professional fields. As professor Ville Pulkki envisions, power plant workers could hear where a possible radiation leak is coming from while divers could benefit from spatial hearing (Zaki, 2024), an ability normally non-functional in water.

With voice recognition, speech synthesis, context-aware sensors, active communication with other devices, and access to online data and large language models, AAR devices have the potential to be the ultimate ubiquitous platforms in helping people's everyday life. However, with an audio-only display, the browsing through information is difficult, so the information should be curated to be as optimal and unbostructive as possible. Through acquiring a deep understanding of the user's situation and environment, the AAR system should 'deliver the right information, at the right time without requiring us to change our focus' (Breunig, 2024).

10.7 Augmented environments

Augmenting the real environment with virtual sounds is a powerful narrative tool: it forces us to observe the immediate reality as a transformed one, yet still real and tangible. This strength can also be utilised in simulations, which—like many stories—ask the question 'What if?' Environment-embedded sounds can be used, for example, to simulate future soundscapes (see Hong et al., 2017). Like the protagonists in our story, designers and citizens could embark on an AAR audio walk and hear the impact of future human actions on the natural and habitant environments.

The immediate space around the user can be augmented for endless applications. In music therapy targeted at motor skill and sensory development, a therapist could use virtual, acousmatic instruments placed around the patient to train coordination, movements and hearing (see Yang, Barde and Billinghurst, 2022). Music and sound mixing engineers will likely expand the use of software that augments virtual loudspeakers in their studio—or bedroom—to enable checking multi-channel mixes without the physical loudspeakers. When more and more music, films, and video games are released in immersive formats, this will become an important convenience in the trade.

10.8 Standards and codecs

In the story, the protagonists receive the interactive audio feed from the museum's server. For this to be possible, there should be standards in place for the delivery of 6DoF virtual audio to the end user's device regardless of its manufacturer and make. The user's pose tracking and other sensory data should, in turn, travel back to the system and be compatible as well.

The new MPEG-I standard is one attempt towards that vision (See Chapter 7.8). Such standards should also streamline the design and production process, lowering the threshold for content creation and making the content available for multiple platforms. This would, however, require that software and hardware manufacturers agree to use a common standard, which is obviously much to ask if any of them feel that the standard cannot deliver high enough quality, or they would rather use their proprietary systems for business or other reasons.

For ubiquitous AAR, seamless connectivity to online services such as map and public transportation data, cloud computing, real-time event and weather data, and other users' shared content would require open APIs and standardised protocols. In our story, the protagonists talked to each other while cycling through a radio link between the earbuds. Later, one shared their urban soundscape simulation with the first-person character. These would also require agreed communication standards, similar to IVAS and MPEG-I.

10.9 Challenges and risks

Money plays a very important role in shaping the future of AAR. For the medium to develop narratively, it is important that AAR experiences are made and explored in museums, cultural heritage sites, theatres, and anywhere where there is a tradition and a need to convey narratives to audiences. However, these are also the places that have the least money. When planning the next budget cycle, investing in an AAR experience or application might not be the top priority. Or, if the technical quality is unconvincing and the storytelling is cumbersome—which is natural for such a young medium—it may remain their last AAR experience for some time.

Some content creators express disappointment, arguing that the high quality of the experience or the externalisation of virtual sounds often goes unnoticed by most people. This perspective is concerning because the details truly matter. Even if individuals lack the vocabulary to articulate how well a virtual character's voice is spatialised, they certainly know whether the character is relatable and if it successfully engages the listener in the story.

For AR devices to work properly, and be a functional part of the ubiquitous computing ecosystem, they need to gather a lot of information about the user's physical location, pose and behaviour, record and analyse people's voices and environmental soundscapes, transfer and process messages, information, often private and sensible. Most of this happens without the user or the people around being aware. After the buzz caused by *Google Glass* and its ability to record video without people noticing, Meta added a small light into their *Ray-Ban Meta* smart glasses to indicate when the video camera is filming and when surround sound is being recorded through the five integrated microphones.

Most of the captured data is probably analysed internally within the devices and deleted after use to save storage space. However, with a constant internet connection, justified for cloud computing needs or 'better user experience', the user has no control of what data is sent to the service provider. This 'fine-grained personal data' (Turner, 2022) is worth gold for those interested in consumer behaviour and targeted marketing, easily slipping away to third parties. Additionally, automatic screening processes may flag unusual but innocent behaviour, causing issues for the user.

'Digital deception', such as deepfake audio content, is another real societal risk, highlighted by the fact that AR technologies can bring these fake objects into the physical world and make them indistinguishable from reality (Turner, 2022). With technology that could erase and replace real-world sounds, a slightly daunting foresight would be a situation where the user of a pair of AAR earbuds is following an election debate and strangely enough, the voice of one of the candidates would always appear muddy and hard to understand while others would sound crystal clear. Yet, verifying the authenticity of the content with AAR would be as simple as taking off the headphones to hear reality for what it is. In contrast, identifying deepfake content on the internet or television would be much more challenging.

10.10 AAR games

Finally, from daunting challenges to uplifting moments, mobile AAR games running on a user's own device hold significant potential, combining several interesting possibilities of AAR. For instance, the multi-player game depicted at the end of our story would be entirely ubiquitous, seamlessly adapting its content to the user's environment. The narrative would unfold through virtual audio attached to real-world locations, transforming ordinary settings into secret hideouts for spies. Sound design elements and narrative content could be partially AI-generated and partially co-authored by the players. A unique twist in this particular game would be that secret audio messages left by players would not be geotagged. Instead, they would be semantically tagged. For example, if a player in another city leaves an item behind a bus stop, another player would find it behind a bus stop, even if it is thousands of kilometres away. While such a concept could be realised with visual AR, it becomes particularly compelling in AAR, where the 'secret' acoustic layer transforms reality into an adventure

10.11 Conclusion

With AAR, embedding auditory information and narratives into the real environment transforms how we perceive and interpret the spaces and objects around us. Rather than simply overlaying sound, the auditory layers actively reshape our understanding of the physical world. It's not the sounds that captivate us—it's the merge.

We are witnessing a trend towards immersive and personalised audio which enables a 'dynamic sound world of our choice' (Kadel, 2024). AAR with an eyes-free, intuitive interface is likely to emerge as a key technology in connecting our lives to the 'augmented space', where real-time data integrates seamlessly into our daily routines (see Manovich, 2003). AAR can enhance and reshape our auditory perception of surrounding events, fostering social interaction while preserving our hearing for treasured unmediated moments.

While users would undoubtedly benefit from holistically designed AAR technology and the rich content envisioned in this book, a content creator would have their own wishlist for the future: streamlined work-flows and intuitive tools that allow anyone to start crafting their own content and exploring the medium. At the same time, professionals could leverage advanced tools to develop even more innovative and compelling applications and experiences—together making audio augmented reality a part of everyone's reality.

References

Albrecht, R., Väänänen, R. and Lokki, T. (2016) 'Guided by Music: Pedestrian and Cyclist Navigation with Route and Beacon Guidance', *Personal and Ubiquitous Computing*, 20 (1), pp. 121–145. Available at: https://doi.org/10.1007/s00779-016-0906-z.

Andre, E. (2024) 'Should I Wear My Hearing Aids All the Time?' Available at: http s://www.earhealth.co.nz/should-i-wear-my-hearing-aids-all-the-time/ (Accessed: 11 December 2024).

Brandenburg, K. et al. (2023) 'Implementation of and Application Scenarios for Plausible Immersive Audio via Headphones', in *Audio Engineering Society Convention* 155, Audio Engineering Society. Available at: https://www.aes. org/e-lib/browse.cfm?elib=22308 (Accessed: 25 November 2023).

Best, V. et al. (2020) 'Sound Externalization: A Review of Recent Research', *Trends in Hearing*, 24, p. 2331216520948390. Available at: https://doi.org/10. 1177/2331216520948390.

Breunig, D. (2024) 'The Rise of Audio Augmented Reality', *Medium*. Available at: https://medium.com/design-bootcamp/the-rise-of-audio-augmented-reality-c56a6 348ff59 (Accessed: 20 November 2024).

Bruhn, S., Multrus, M. and Varga, I. (2022) 'IVAS – Taking 3GPP Voice and Audio Services to a New Immersive Level', *3GPP Highlights*, October, pp. 8–9.

Das, A.-S. (2023) 'How the Apple Vision Pro Will Shift the Audio Experience: Part 2', *Fansea*, 11 October. Available at: https://medium.com/fansea/how-the-apple-vi sion-pro-will-shift-the-audio-experience-part-2-b5788051ebaf (Accessed: 11 December 2024).

Duong, N.Q.K. et al. (2017) 'Audio Zoom for Smartphones Based on Multiple Adaptive Beamformers', in P. Tichavský et al. (eds) *Latent Variable Analysis and Signal Separation*. Cham: Springer International Publishing, pp. 121–130. Available at: https://doi.org/10.1007/978-3-319-53547-0_12.

Envision (2023) 'Envision Glasses – Smart Glasses for People Who Are Blind or Low Vision'. Available at: https://www.letsenvision.com/glasses/home (Accessed: 12 December 2024).

Hong, J.Y. et al. (2017) 'Spatial Audio for Soundscape Design: Recording and Reproduction', *Applied Sciences*, 7 (6), p. 627. Available at: https://doi.org/10. 3390/app7060627.

Kadel, O. (2024) 'The Evolution of Spatial Audio in Immersive Storytelling', *MPSE Wavelength*, June, pp. 72–83.

Kiss, F., Mayer, S. and Schwind, V. (2020) 'Audio VR: Did Video Kill the Radio Star?', *Interactions*, 27, pp. 46–51. Available at: https://doi.org/10.1145/3386385.

Kucinski, W. (2018) 'A-10C Pilots Will Get 3D-Audio to Increase Situational Awareness'. Available at: https://www.sae.org/site/news/2018/11/a-10c-pilots-will-get-3d-audio-to-increase-situational-awareness (Accessed: 10 October 2024).

Kulawiak, P.R. (2021) 'Academic Benefits of Wearing Noise-Cancelling Headphones During Class for Typically Developing Students and Students with Special Needs: A Scoping Review', Edited by D. Schussler. *Cogent Education*, 8 (1), p. 1957530. Available at: https://doi.org/10.1080/ 2331186X.2021.1957530.

Leaver, A.M. (2024) 'Perceptual and Cognitive Effects of Focal tDCS of Auditory Cortex in Tinnitus'. *medRxiv*, p. 2024.01.31.24302093. Available at: https:// doi.org/10.1101/2024.01.31.24302093.

Lovich, S.N. et al. (2023) 'Parametric Information About Eye Movements Is Sent to the Ears', *Proceedings of the National Academy of Sciences*, 120 (48), p. e2303562120. Available at: https://doi.org/10.1073/pnas.2303562120.

Lylykangas, J. et al. (2023) 'Ears Outside the Car: Evaluation of Binaural Vehicular Sounds and Visual Animations as Driver's Blind Spot Indicators', in *Adjunct Proceedings of the 15th International Conference on Automotive User Interfaces and Interactive Vehicular Applications*. Ingolstadt Germany: ACM, pp. 60–65. Available at: https://doi.org/10.1145/3581961.3609900.

Manovich, L. (2003) 'The Poetics of Augmented Space', in A. Everett and J.T. Caldwell (eds) *New Media: Theories and Practices of Digitextuality*. Routledge, pp. 75–92.

McGill, M. et al. (2020) 'Acoustic Transparency and the Changing Soundscape of Auditory Mixed Reality', in *Proceedings of the 2020 CHI Conference on Human Factors in Computing Systems*. Honolulu, Hawaii, USA: ACM, pp. 1–16. Available at: https://doi.org/10.1145/3313831.3376702.

Meyer-Kahlen, N. (2024) *Transfer-Plausible Acoustics for Augmented Reality*. PhD thesis. Aalto University. Available at: https://urn.fi/URN:ISBN: 978-952-64-1913-8.

Purcher, J. (2024) 'A Top Apple Analyst Predicts that Apple Could Introduce Next-Gen AirPods with an IR Camera, Apple Intelligence & More for 2026', *Patently Apple*. Available at: https://www.patentlyapple.com/2024/07/a-top -apple-analyst-predicts-that-apple-could-introduce-next-gen-airpods-with-a n-ir-camera-apple-intelligence-more-for-202.html (Accessed: 25 October 2024).

Schmalstieg, D. and Hollerer, T. (2016) *Augmented Reality: Principles and Practice*. Addison-Wesley Professional.

Signia (2024) 'Signia Integrated Xperience Hearing Aids – All Models', *Signia*. Available at: https://www.signia.net/en/hearing-aids/integrated-xperience/ (Accessed: 22 October 2024).

Sonova (2024) 'Sonova Announces Two New Hearing Aid Platforms, Including the First Hearing Aid with Real-Time AI to Address Most Pressing Need in Hearing Loss', *Sonova International*. Available at: https://www.sonova.com/ en/investor-news/sonova-announces-two-new-hearing-aid-platforms-inclu ding-first-hearing-aid-real-time (Accessed: 7 August 2024).

Turner, C. (2022) 'Augmented Reality, Augmented Epistemology, and the Real-World Web', *Philosophy & Technology*, 35 (1), p. 19. Available at: https://doi. org/10.1007/s13347-022-00496-5.

Veluri, B. et al. (2023) 'Semantic Hearing: Programming Acoustic Scenes with Binaural Hearables', in *Proceedings of the 36th Annual ACM Symposium on User Interface Software and Technology*. New York, NY, USA: Association for Computing Machinery (UIST '23), pp. 1–15. Available at: https://doi.org/ 10.1145/3586183.3606779.

Veluri, B. et al. (2024) 'Look Once to Hear: Target Speech Hearing with Noisy Examples', in *Proceedings of the 2024 CHI Conference on Human Factors in Computing Systems*. New York, NY, USA: Association for Computing Machinery (CHI '24), pp. 1–16. Available at: https://doi.org/10.1145/3613904.3642057.

Veronese, S.et al. (2023) 'Ten-Year Follow-Up of Auditory Brainstem Implants: From Intra-Operative Electrical Auditory Brainstem Responses to Perceptual Results', *PLOS ONE*, 18 (3), p. e0282261. Available at: https://doi.org/10.1371/journal.pone.0282261.

Yang, J., Barde, A. and Billinghurst, M. (2022) 'Audio Augmented Reality: A Systematic Review of Technologies, Applications, and Future Research Directions', *Journal of the Audio Engineering Society*, 70 (10), pp. 788–809.

Zaki, S. (2024) 'Reportaasi | Harva tietää, että Espoossa piilee "Suomen hiljaisin huone"', *Helsingin Sanomat*. Available at: https://www.hs.fi/pkseutu/art-2000010425471.html (Accessed: 25 May 2024).

Index

Pages in *italics* refer to figures, pages in **bold** refer to tables.

For Product Safety Concerns and Information please contact our EU
representative GPSR@taylorandfrancis.com
Taylor & Francis Verlag GmbH, Kaufingerstraße 24, 80331 München, Germany